AF685767

Tropical and Subtropical Fruits: Flavors, Color, and Health Benefits

ACS SYMPOSIUM SERIES **1129**

Tropical and Subtropical Fruits: Flavors, Color, and Health Benefits

Bhimanagouda S. Patil, Editor
Texas A&M University, College Station, Texas

Guddadarangavvanahally K. Jayaprakasha, Editor
Texas A&M University, College Station, Texas

Coralia Osorio Roa, Editor
Universidad Nacional de Colombia, Bogotá, Colombia

Kanjana Mahattanatawee, Editor
Siam University, Bangkok, Thailand

Sponsored by the
ACS Division of Agricultural and Food Chemistry, Inc.

American Chemical Society, Washington, DC

Distributed in print by Oxford University Press

Library of Congress Cataloging-in-Publication Data

Tropical and subtropical fruits : flavors, color, and health benefits / Bhimanagouda S. Patil, editor, Texas A University, College Station, Texas, Guddadarangavvanahally K. Jayaprakasha, editor, Texas A&M University, College Station, Texas, Coralia Osorio Roa, editor, Universidad Nacional de Colombia, Bogotá, Colombia, Kanjana Mahattanatawee, editor, Siam University, Bangkok, Thailand ; sponsored by the ACS Division of Agricultural and Food Chemistry, Inc.

pages cm. -- (ACS symposium series ; 1129)

Includes bibliographical references and index.

ISBN 978-0-8412-2893-1 (alk. paper)

1. Antioxidants--Health aspects. 2. Tropical fruit--Nutrition. 3. Nutrition. I. Patil, Bhimanagouda S., 1962- editor of compilation. II. American Chemical Society. Division of Agricultural and Food Chemistry, sponsoring body.

RB170.T76 2013

613.2'86--dc23

2013011959

The paper used in this publication meets the minimum requirements of American National Standard for Information Sciences—Permanence of Paper for Printed Library Materials, ANSI Z39.48n1984.

Copyright © 2013 American Chemical Society

Distributed in print by Oxford University Press

All Rights Reserved. Reprographic copying beyond that permitted by Sections 107 or 108 of the U.S. Copyright Act is allowed for internal use only, provided that a per-chapter fee of $40.25 plus $0.75 per page is paid to the Copyright Clearance Center, Inc., 222 Rosewood Drive, Danvers, MA 01923, USA. Republication or reproduction for sale of pages in this book is permitted only under license from ACS. Direct these and other permission requests to ACS Copyright Office, Publications Division, 1155 16th Street, N.W., Washington, DC 20036.

The citation of trade names and/or names of manufacturers in this publication is not to be construed as an endorsement or as approval by ACS of the commercial products or services referenced herein; nor should the mere reference herein to any drawing, specification, chemical process, or other data be regarded as a license or as a conveyance of any right or permission to the holder, reader, or any other person or corporation, to manufacture, reproduce, use, or sell any patented invention or copyrighted work that may in any way be related thereto. Registered names, trademarks, etc., used in this publication, even without specific indication thereof, are not to be considered unprotected by law.

PRINTED IN THE UNITED STATES OF AMERICA

Foreword

The ACS Symposium Series was first published in 1974 to provide a mechanism for publishing symposia quickly in book form. The purpose of the series is to publish timely, comprehensive books developed from the ACS sponsored symposia based on current scientific research. Occasionally, books are developed from symposia sponsored by other organizations when the topic is of keen interest to the chemistry audience.

Before agreeing to publish a book, the proposed table of contents is reviewed for appropriate and comprehensive coverage and for interest to the audience. Some papers may be excluded to better focus the book; others may be added to provide comprehensiveness. When appropriate, overview or introductory chapters are added. Drafts of chapters are peer-reviewed prior to final acceptance or rejection, and manuscripts are prepared in camera-ready format.

As a rule, only original research papers and original review papers are included in the volumes. Verbatim reproductions of previous published papers are not accepted.

ACS Books Department

Contents

Indexes

Preface

Fruits are one of the most important links in the food chains essential for sustenance of life on the planet Earth and can be classified as tropical or subtropical, depending on where they are cultivated. Asia and the Pacific Rim are the major producing regions, followed by Latin America, the Caribbean, and Africa. Most developing countries have a rich biodiversity, with a significant number of species that are widely consumed among native people but completely unknown in the rest of the world. These fruits exhibit exotic characteristics due to their adaptation to special climatic conditions, this fact make them attractive for new markets. For example, fruits from South America, such as Cape gooseberry (*Physalis peruviana*), passion fruit (*Passiflora edulis* Sims), pitahaya (*Hylocereous triangularis*), açai (*Euterpe oleracea*), and tamarillo (*Solanum betaceum* Cav.) have been commercialized in Europe and are used in "haute cuisine". Likewise, mangosteen (*Garcinia mangostana*), durian (*Durio zibethinus*), lychee (*Litchi chinensis*), and rambutan (*Nephellium lappaceum*) are representative of the Asian Pacific Rim; also, jackfruit (*Artocarpus heterophyllus*), and carambola (*Averhoa carambola*) are cultivated in India and surroundings. These fruits not only satisfy hunger but also provide year-round sources of nutrients and exhibit a distinctive flavor and appearance that implies good consumer acceptance. Each of these fruits has a characteristic volatile composition that creates unique and distinctive volatile flavour profiles. Non-volatile compounds, such as polyphenols, peptides, or terpenoids, among others, are responsible for fruit colour and health benefits. Fruit color is a key indicator of ripeness and the first sensory parameter that captures the consumer's attention. The vast genetic diversity in tropical and subtropical fruits makes them nature's bounty of bioactive compounds with flavor, color and health benefits.

This book evolved from a symposium, "Tropical and Subtropical Fruits: Flavors, Color, and Health Benefits," held at the 242th annual meeting of the American Chemical Society on August 28–September 1, 2011 at Denver, CO and hosted by the Division of Agricultural & Food Chemistry. Subtropical fruits have been studied for many years; in contrast, the number of publications related to sensory and biofunctional properties of tropical fruits has only recently increased. Therefore, the objective of this symposium was to provide an update on the state of the art in this regard, and to highlight the role of fruits as source of bioactive compounds with health benefits. At this meeting, 21 national and international scientists from ten different countries were successfully brought together to exchange ideas. This symposium was aimed at providing an avenue for scientists around the world to share and gain innovative ideas in this emerging area of research, and also provided an opportunity for researchers to widen their horizons of chemistry and initiate interdisciplinary collaborations. Thus, it was possible to exchange of insights among researchers in distinct branches of science such as horticulture, food and natural product chemistry, agronomy and others. Apart from academia, this symposium also benefited the industrial community by providing an ideal platform to address the specific concerns and scientific needs of industrial science. For example, the demand for natural flavors and colors has increased tremendously, based on an increasing interest in natural products, coupled with the recent implications of bioactive compounds in health benefits.

Multiple inter-disciplinary strategies, techniques and approaches are being pursued globally to define and harness the beneficial attributes of these varied fruit species. These attributes, including flavor, color, and health benefits, are substantially affected by plant species, genotype, environmental conditions, and physiological changes. This dynamic area of research requires both international and scientific collaboration. The 2011 Tropical and Subtropical Fruits: Flavors, Color, and Health Benefits Symposium and the resulting book aim to potentiate collaboration and the exchange of ideas in this important, emerging field of research, particularly in the examination of plant bioactive compounds.

Among the bioactive phytochemicals, polyphenols stand out as responsible for many bioactivities. The polyphenols are plant secondary metabolites involved in defense against ultraviolet radiation or attack by pathogens, and can be classified, based on chemical structure, as phenolic acids, flavonoids, stilbenes, and lignans. The flavonoids, consisting of two aromatic rings (A and B) that are bound together by three carbons that form an oxygenated heterocycle (ring C), may themselves be divided into six subclasses based on the type of heterocycle involved: flavonols, flavones, isoflavones, flavanones, anthocyanidins, and flavanols (catechins and proanthocyanidins).

As detailed in this book, much of the bioactivity of some fruits can be explained by the presence of different classes of polyphenols. For example, piceatannol, a stilbene isolated from of passion fruit seeds showed anti-aging effects, thus transforming a waste fruit product into a promising resource for the development of cosmetic/medicinal products. The most remarkable bioactivity of fruits is antioxidant activity and this book also provides several examples of anthocyanin-rich fruits, including pomegranate (*Punica granatum* L.) and five wild Colombian fruits. Additionally, other flavonoids, such as dihydrochalcones, flavones, and flavanones in kumquat (*Citrus japonica*), and procyanidins in cherimoya (*Annona cherimola*), also are responsible for the antioxidant activity and cytoprotective activity.

The first topic examined in this volume is the chemistry of tropical and subtropical fruits, with studies involving isolation of volatile aroma constituents and their identification from citrus, pineapple, and date palm fruits, as well as plant breeding and genetic approaches to improve flavor and nutrients. The second topic examines the effect of climate, salinity and genetic factors on health-promoting and color properties in pomegranates, high hydrostatic pressure processing as a strategy to increase carotenoids of tropical fruits, and preparative separation and pigments from *Opuntia ficus-indica* by ion-pair high-speed countercurrent chromatography. Finally, the book examines health-promoting properties of citrus volatile constituents, *Moringa oleifera,* and passion fruit, discusses the antioxidant properties of the anthocyanin-rich fruits annona and citrus and shows the effects of acai berries and walnuts on brain health.

The editors acknowledge and thank all authors for their patience, hard work, and timely contributions, as well as the rest of the speakers who gave oral presentations in this symposium. Further, we wish to thank deeply the reviewers for their critical suggestions and comments, which made this book possible. Finally, we would like to thank ACS Books Division, Bob Hauserman, Tim Marney, Kat Squibb as well as individuals from the press, especially Mary Calvert and Pamela Kame, for publishing this symposium series.

Bhimanagouda S. Patil
Texas A&M University
College Station, Texas

G. K. Jayaprakasha
Texas A&M University
College Station, Texas

Coralia Osorio Roa
Universidad Nacional de Colombia
Bogotá, Colombia

Kanjana Mahattanatawee
Siam University
Bangkok, Thailand

Flavor and Color in Fruits: Separation and Processing

Chapter 1

Preparative Separation and Pigment Profiling of Betalains from Fruits of *Opuntia ficus* by Ion-Pair High-Speed Countercurrent Chromatography (IP-HSCCC) and Off-Line LC-ESI-MS/MS

Gerold Jerz,[*,1] Bettina Klose,[1] Sławomir Wybraniec,[2] Socorro Villanueva Rodriguez,[3] and Peter Winterhalter[1]

[1]Institute of Food Chemistry, Technische Universität Braunschweig, Schleinitzstrasse 20, 38106 Braunschweig, Germany
[2]Department of Analytical Chemistry, Institute C-1, Faculty of Chemical Engineering and Technology, Cracow University of Technology, ul. Warszawska 24, Cracow 31-155, Poland
[3]Centro de Investigación y Asistencia en Technología y Diseño del Estado Jalisco, Guadalajara, Mexico
[*]E-mail: g.jerz@tu-bs.de. Fax: +49-531-391-7230.

Opuntia spp. (prickly pear) cultivation is playing a steadily growing role as a drought tolerant plant resource for fruit, and fodder production and also land conservation worldwide. From the view point of global climatical changes and the increasing needs for a consistent and reliable fruit production, cacti fruits are almost ideal crops for areas endangered by desertification and inconsistent precipitation levels. Various cacti plants have their origin from arid regions of North-America, and today they are distributed throughout most of all dry sub-tropical and tropical areas of the world. Betalain pigment profilings of betacyanins with their respective degradation products and betaxanthins occurring in yellow, orange and violet *Opuntia* fruits from Sicilia were done by using two chromatographic dimensions. Fractions recovered by preparative ion-pair HSCCC were investigated by *off-line* LC-ESI-MS/MS analysis.

© 2013 American Chemical Society

Betalains are naturally occurring pigments which can be divided into red-purple betacyanins and yellow-orange colored betaxanthins. They are present in some fruits and vegetables of specific plant families used for human nutrition such as the family Chenopodiaceae with *Beta vulgaris* – red and orange beets (*1–3*), swiss chards (*4*), *Amaranthus tricolor* from the family Amarathaceae (*5*), and the Andean tuber crop ulluco (*Ullucus tuberosus*) from the family Basellaceae (*6*). From the tropical and sub-tropical origin, the family of Cactaceae is providing a huge variety of edible cacti fruits such as *Hylocereus polyrhizus* (pitahaya) (*7*), *Opuntia ficus indica* (tuna) (*2, 3*), and other *Opuntia* spp, which could be used for food production. Formerly, *Opuntia ficus* fruits, and leafy green cladodes were the "food for the poor" but today the scientific focus moved to the existing nutritional values and potential health properties. Already 3000-8000 years ago, Aztec people in Mexico used *Opuntia* spp. Beside the nutritional values, the fruits contain high amounts of antioxidant compounds such as glutathione and betalains which seem to be still active in blood plasma.

Less known cacti fruits with small production sizes from wild collections are *Myrtillocactus geometrizans* (garambullo) (*8, 9*), and *Stenocereus queretoensis* (pitaya) (*10*). Systematic cultivation of these cacti fruits could be increased for future production.

Today, red beet and also increasing amounts of cacti pigment extracts e.g. from *H. polyrhizus* are used in various dairy products, such as yoghurts and ice creams due to their favorably good coloring properties and stabilities at almost neutral pH-value. In this case, the red-violet color hues of betalains cannot be achieved or substituted by anthocyanin extracts recovered from fruit and vegetable sources. Interestingly, in comparison to the overwhelming amount of scientific investigations on positive health effects of red-violet colored anthocyanins in the human diet, the knowledge about influences and specific functionalities of betalains in human nutrition is rather limited. In recent literature, some promising studies had shown positive aspects for the consumption of *Opuntia ficus* specifically related to the antioxidant pigments betanin and indicaxanthin. Induction of an improved antioxidative status in human red blood cells was observed (*11–16*). In the future, additional investigations will be required to understand aspects of the physiology of betalains in the human body, possible chemopreventive actions and benefits of a diet rich in betalain pigments. Unfortunately, processing of betalain pigments is extremely difficult. Violet betacyanins, and also their red colored first stage degradation products – the decarboxy-betacyanins (*1*) - furthermore yellow or orange colored betaxanthins are highly instable and require gentle processing conditions to preserve the intense coloring properties during thermal pasteurization. Isolations of pure standards of betacyanins and betaxanthins by preparative lab-scale procedures with the aim of doing biological evaluations are also hampered due to low compound stabilities and result in low yields of reference compounds. In crude extracts, betalain pigments are still protected by the antioxidant environment of the polyphenolic matrix acting as radical scavengers. Clean-up procedures with the result of improved pigment purities simultaneously lead to an enormously increased compound instability. Reactivity and pigment deterioration in the laboratory can be highly influenced by sun-light, oxygen impact, contact to certain organic solvents such as alcohols. From aspects of chromatography, solid phase materials e.g. silica gel, partly also reversed phase C18 material with endcapping properties, organic size exclusing resins (Sephadex LH20) or organic cation-exchanger (Amberlite XAD-7, XAD-17) had caused rapid pigment degradation observed from our personal experience.

As shown in ealier studies, *ion-pair high-speed countercurrent chromatography* (IP-HSCCC) as a complete support-free, all liquid chromatographic technique can be effectively used for *preparative* scale isolation procedures for betalain recovery (*17–20*). The mode of separation during *countercurrent chromatography* (CCC) can be explained by fast mixing and settling processes of the sample analytes in biphasic solvent systems induced by rapidly changing centrifugal force fields which will be generated by a so-called *J*-type centrifuge with serially connected coil columns rotating in a planetary-type system. Separation columns are very long Teflon tubes wrapped around these rotating coil-bobbins. Further detailed information about CCC techniques is available in the literature (*21*, *22*).

Betalains are very polar compounds, and the used HSCCC-devices operated with biphasic solvent systems omit completely undesired chemisorptive effects of analytes to polar solid phase materials. A large number of already evaluated solvent mixtures generating two-phase systems can be applied for HSCCC (*21*, *22*), but for separations of highly instable and polar betalain structures, suitable solvent systems not inducing a fast degradation of these pigments are limited. Standard systems with low impact to pigments consist of *t*-butylmethylether (TBMe), *n*-butanol, acetonitrile and water (*18*). Different perfluorinated carboxylic acids were used as ion-pair modifiers to move pigments from the so called aqueous mobile phase to the organic stationary phase and guarantee a good partition ratio K_D for the compounds. In a pilot study a freshly prepared *Opuntia ficus* juice from orange *Opuntia* was evaluated by different perfluoro acid additives on IP-HSCCC to optimize chromatography. For later larger scale pigment fractionation, C18-reversed phase pre-cleaned pigment extracts of *Opuntia ficus* fruits were used for IP-HSCCC injection.

In most of the IP-HSCCC experiments on *Opuntia* pigment extracts the so-called *elution-extrusion* CCC experimental approach or two-colum volume method was used (*23*) and recovered the complete pigment profile of the injected sample. Minor concentrated lipophilic betacyanins such as 15*R*- and 15*S*-phyllocactin and a feruloyl-substituted betacyanin were identified in violett *Opuntia* fruits. The recovered HSCCC fractions were investigated by using a second chromatographic dimension of C18 HPLC and ESI-MS/MS analysis and revealed a highly complex metabolite profile in each fraction not necessarily related to pigment structures.

Experimental Procedure

Extraction Process of Betacyanin and Betaxanthin Pigments

Investigated *Opuntia ficus* fruits were of yellow, orange and purple colors and were imported from Sicily (Italy). An amount of 500 g fruits of each color variety were manually depealed, the flesh with seeds was worked up quickly to prevent alterations in the pigment profiles. Fresh juices were prepared using a household blender separating the juice from the seeds and solid tissue materials. For prevention of pigment decomposition of sensitive betalains, the juices (12.5-12.7 Brix, pH 6.0) were poured into a mixture of water and acetonitrile which had been acidified with TFA (1%) until a final ratio of 4 % organic solvent in the fruit liquids (total volume approx. 1L). Ascorbic acid was added for antioxidant stabilization. Then the juices were filtered through Büchner funnels and filtrates were immediately frozen (-25 °C), and directly lyophilized (Christ Freeze Dryer beta 2-8 LD plus, Osterode, Germany: cooler -90 °C; vacuum pump: Chemistry Hybrid Pump RC-6, Vaccubrand, Wertheim, Germany: vacuum 0.05 mbar) using a nitrogen cooled high capacity solvent trap (approx. 3 L) for vacuum pump protection. Perfluoro acids used for the extractions and IP-HSCCC were purchased from Aldrich (Steinheim, Germany) in p.a. quality: trifluoro-acetic acid (TFA), perfluoro-glutaric acid (PFGA), pentafluoro-propionic acid (PFPA), heptafluoro-butyric acid (HFBA). Solvents used for HSCCC (TBMe, acetonitrile, n-butanol) were of analytical grade and were purchased from Sigma-Aldrich (Steinheim, Germany), acetonitrile for LC-ESI-MS from Honeywell Speciality Chemicals (Seelze, Germany), and Nanopure water was generated by a laboratory clean water unit (Werner Reinstwasser System, diamond analytic, Leverkusen, Germany).

C18-Reversed Phase Middle Pressure Chromatography (MPLC) Clean-Up of *Opuntia ficus* Pigment Extracts

For clean-up of the polysaccharide rich *Opuntia ficus* dried crude pigment extracts, a C18-MPLC column (Büchi, Switzerland) (7 cm i.d., length 25 cm) was filled with C18-reversed phase material (Figure 1).

Sequentially, these dried juices of *Opuntia ficus* color varieties yellow, orange, purple were cleaned by this procedure by dissolving in Nanopure water, TFA (0.7% v/v) and loaded to the column. The pigments were trapped by the aqueous TFA to the C18 solid phase and subsequent thorough rinsing with the mixture eluted sugars, amino acids and fruit acids. For betalain elution, a mixture of acetonitrile (~50%), nanopure water, and TFA was used. The recovered pigment eluates containing acetonitrile were directly freeze-dried for IP-HSCCC (Figure 1). Unfortunately, the use of perfluoro-acids for extraction and the later IP-HSCCC separation will complicate *in-vitro* or *in-vivo* biological evaluations. The formation of perfluoro salts of betalains will occur and the products definitely require a thorough clean-up by C18 reversed phase preparative HPLC using acetic acid or formic acid solvent additives. HPLC separation and lyophilization of the pigments should liberate the compounds from the toxic perfluoro acids. However, each additional chromatography step could possibly reduce the final yields as described above due to known pigment instabilities.

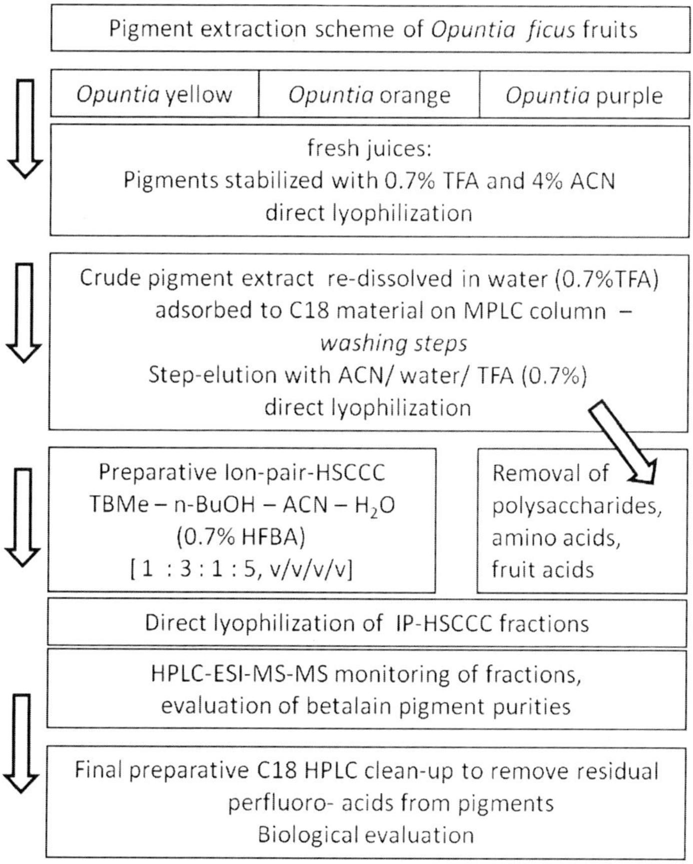

Figure 1. Extraction, fractionation and isolation flow scheme of betacyanin and betaxanthin pigments from dried juices (crude pigment extracts) of colored Opuntia fruits.

Experimental Set-Up of the IP-HSCCC Apparatus and the Use of Different Ion-Pair Reagents

The preparative HSCCC instrument was a multilayer coil planet centrifuge model CCC 1000 (Pharma-Tech Research Corp., U.S.A.), equipped with three preparative coil columns connected in series (polytetrafluorethylene (PTFE) tubing-volume: 850 mL). A manual sample injection valve with a 25 mL loop was used to introduce the sample into the coil column.

The separation of indicaxanthin from orange *Opuntia* fruits was evaluated by the observed retention time shifts caused by variation of different perfluoro acid additions. The biphasic HSCCC solvent system TBMe – n-BuOH – ACN – H_2O was used with varying additions of perfluoro acid reagents such as perfluoro-glutaric acid 0.2% (v/v), and 1.0% (v/v) for TFA, PFPA or HFBA to the aqueous phase, respectively (Figure 2).

The biphasic solvent systems were freshly prepared before the IP-HSCCC experiments and divided in a separatory funnel with total amounts of 2.5 L of each phase.

The mobile phase was delivered with a Biotronik BT 3020 HPLC pump (Jasco, Gross-Umstadt, Germany) (for details of the HSCCC apparatus cf. (*24*)). The CCC equipment was operated in the '*head-to-tail*' mode using the upper solvent layer as the stationary phase. The mobile phase was pumped during *elution mode* at a flow rate of 3.0 mL/ min. Velocity of the HSCCC apparatus was set to 850 rpm. The injection was done with a low pressure injection valve (Rheodyne, USA) and a 25 mL sample loop. The samples were dissolved in aliquots of both phases and filtered before injection. For recovery of the whole polarity window of the injected sample, certain HSCCC runs (Figure 2) were additionally operated in the *extrusion mode* to extrude the already fractionated compounds with high affinity to the stationary phase from the CCC centrifuge to the fraction collector. The *elution-extrusion* HSCCC approach, or two column volume ($2V_C$)-methodology was suggested by Berthod et al. (*2*).

Pigment elution during the IP-HSCCC chromatographic runs was measured at λ 440 and 540 nm with two serially connected Knauer UV-vis detectors equipped with large preparative flow cells (Berlin, Germany)

In the preparative experiments for investigation of retention time shifts of indicaxanthin (Figure 2) injection amounts between 900-990 mg were applied. These HSCCC solvent systems (TBMe – n-butanol – ACN – water) had shown significant lower pigment alterations during the preparative fractionation and isolation of betalains including the subsequent drying process (*17–20*). The collected HSCCC fractions from the chromatographic *elution* and *extrusion*, were combined, directly frozen and lyophilized not to cause rapid pigment degradation prior to LC-ESI-MSMS analysis

For the IP-HSCCC separation of C18 MPLC pre-cleaned *Opuntia* pigment extracts the amount of HFBA was reduced to 0.7% in the TBMe – n-butanol – acetonitrile – water system. The fractionation of pigments was done in case of the betacyanins by their changing red-violet color hues in the tubes, and for indicaxanthin by the chromatogram at λ 440 nm. Yields from the IP-HSCCC runs of C18-pre-cleaned *Opuntia ficus* pigment extracts (Figure 2) operated with 0.7% (v/v) HFBA were as follows.

Total injection amount 1.02 g (purple *Opuntia* extract). Recovered fractions: F1, 91 mg; F2, 73 mg; F3, 61 mg; F4, 31 mg; F5, 22 mg; F6, 18 mg; F7, 17 mg; F8, 14 mg; F9, 19 mg; F10, 38 mg; F11, 18 mg; F12, 13 mg; F13, 22 mg; F14, 17 mg; F15, 9 mg; F16, 8 mg; F17, 7 mg; F18, 9 mg, F19, 26 mg; F20, 39 mg; F21, 60 mg; F22, 25 mg; F23, 21 mg; F24, 8 mg; F25, 85 mg; F26, 12 mg; F27, 34 mg; F28, 48 mg; F29, 35 mg. Σ recovery: 880 mg = 86.3 %. F1-F18: fractions with high betacyanin contents; F21-F22: the fractions with high content of indicaxanthin resulted in a yield of 85 mg (8 %).

Yields for IP-HSCCC (orange C18 cleaned material): total injection amount 0.98 g with recovered purified indicaxanthin fractions F12-F15: 135 mg (13.7%). Yields for IP-HSCCC (yellow C18 cleaned material): total injection amount 0.85 g with recovered purified indixanthin fractions F14-F17: 110 mg (12.9%).

Pigment Profiling by LC-ESI-MS-MS

An HPLC system (pump 1100 series, autosampler 1200 series) from Agilent Technologies (Böblingen, Germany) was coupled with an ESQUIRE LC ESI-MS/MS ion-trap system from Bruker Daltonics (Bremen, Germany).

LC-Conditions and Gradient Program for ESI-MS-MS

For C18 HPLC a ProntoSil C18Aq column (5 μm, 250 x 4.6 mm, Knauer, Berlin, Germany) with a flow rate of 0.8 mL/min was used. Solvent A: Nanopure water/ formic acid 98 : 2 (v/v), solvent B: acetonitrile/ formic acid 98 : 2 (v/v). The HPLC-separations were carried out at ambient temperature with following conditions: 0 min (1 % B), 20 min (20 % B), 35 min (50 % B), 45 min (100 % B), 55 min (100 % B), 60 min (1 % B), 65 min (1 % B).

ESI-MS/MS Conditions

ESI-MS-MS parameter settings: drying gas was nitrogen (flow 12.0 L/min, 320 °C), and nebulizer pressure was set to 40 psi. ESI-MS/MS ionization parameters (pos. mode): capillary 3500 V, end plate off set -500 V, capillary exit -113 V, trap drive 55.0, target mass range *m/z* 500, compound stability 100%, trap drive level 120%, ICC target 50000, max. accumulation time 200 ms, charge control on, scan range m/z 50-2200, threshold auto MS/MS 500, MS/MS experiments afforded a fragmentation amplitude value of 1.2 V.

Results and Discussion

Variation of Perfluoro Acid Ion-Pair Reagents for Recovery of Indicaxanthin from Orange *Opuntia ficus* Fruits by IP-HSCCC

The effects of using ion-pair forming perfluoro acids on betaxanthin retention time shifts during HSCCC was evaluated by using reagents with increasing lipophilicity. Similar injection amounts of the crude pigment extract from orange *Opuntia* fruits were used (900-990 mg) to maintain similar experimental conditions. In this pilot study, crude extract (dried juice) was not pre-cleaned by C18 MPLC chromatography. As seen in Figure 2, the principal betaxanthin of *Opuntia* fruits is indicaxanthin and was eluted in system A after 100 min measured from the begin of mobile phase 'break-through' of the HSCCC run (*21*, *22*). Replacing the non-volatile perfluoro glutaric acid (PFGA) by more lipophilic reagents TFA, PFPA, and HFBA, the observed retention time shifts (Rt) for indicaxanthin were strongly enlarged to the maximum time of 280 min in the case of using HFBA additive. In general it could be mentioned that the use of HFBA produced indicaxanthin of higher purity due to a longer retention time on HSCCC.

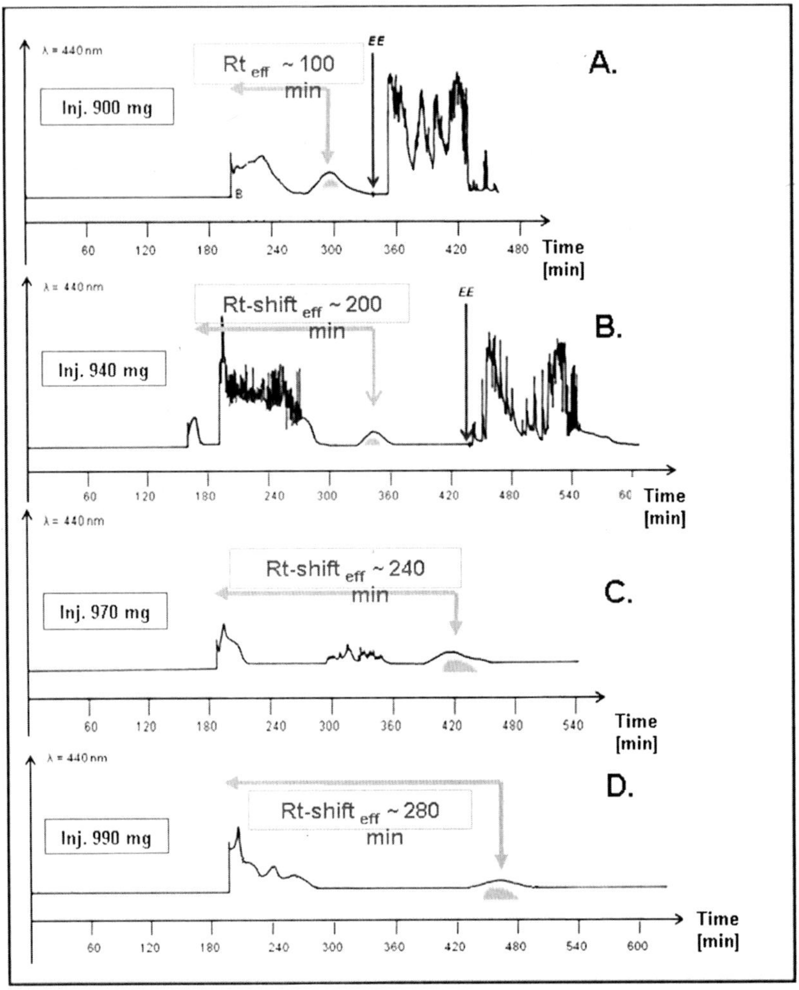

Figure 2. The IP-HSCCC chromatography of crude pigment extract of orange pigmented Opuntia fruits with TBMe – n-BuOH – ACN – H_2O [1:3:1:5, v/v/v/v]. Separation A: 0.2% PFGA; separation B:1.0% TFA; separation C: 1.0% PFPA; separation D: 1.0% HFBA. Injection amounts 900-990 mg. Retention time shifts (Rt) of indicaxanthin induced by increasing lipophilicity of acids (PFGA→TFA→PFPA→HFBA).

As expected from the pigment structures, the perfluoro acids induced much stronger retention time shifts for betaxanthins than for betacyanins. So far no betaxanthins are known with glycosidation, but betacyanins occur in general with sugar substitution, in minimum one glucose-moiety (15*S*/15*R*-betanin or gomphrenin). Beside the ion-pair masking effect and interaction of the dissociated and negatively charged perfluoro acids with the positive charge of the diazaheptamethinium betalain partial structure, also hydrogen bond formations of the perfluoro acids with the carboxyl-group(s) of the betalain pigments are induced and resulted in longer retention times (Figure 3).

The separations of *Opuntia* crude pigment extracts yielded pure indicaxanthin in very low amounts of (~2-3 mg). Therefore, a pre-cleaning step prior to IP-HSCCC is advised to achieve a significant increase of pigments for the preparative HSCCC. TFA is available for a reasonable price and could be used for rapid indicaxanthin isolation. For conducting biological assays a further clean-up of indicaxanthin by preparative C18 HPLC and formic or glacial acid as gradient additive is definitely required to remove the perfluoro acids which are forming salts with the betalain pigments after IP-HSCCC (Figure 1).

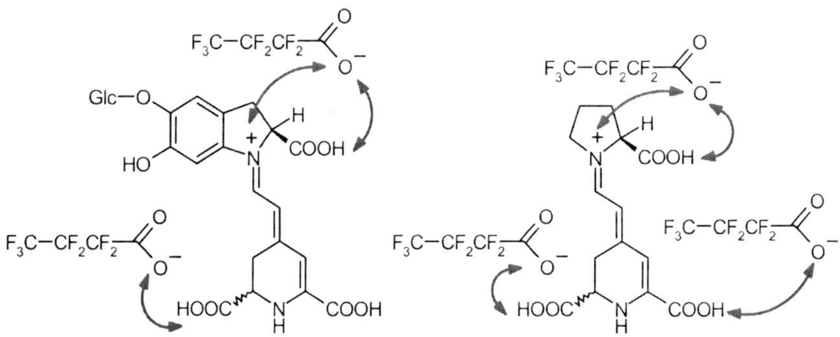

Figure 3. Mechanisms of ion-pair formation in betalain structures during IP-HSCCC: red arrow – charge masking by ion-pair formation, blue arrow – postulated hydrogen bond interaction of perfluoro acids (e.g. HFBA) to carboxylic groups. Left structure: 15S/15R-betanin, right structure: 11S/11R-indicaxanthin.

Metabolite and Pigment Profiles of Pigmented *Opuntia* Fruits (*Opuntia ficus, Opuntia ayrampo*) by LC-ESI-MS/MS

LC-ESI-DAD-MS pigment profiles of 10 different *Opuntia* cultivars from Mexico were previously reported (*25*), and also from Sizily (*2*, *3*). For evaluation of the different pigment profiles of yellow, orange and purple *Opuntia ficus* fruits, LC-ESI-MS/MS analysis were conducted after pre-cleaning/enrichment of crude extracts on a C18 reversed phase column (cf. Experimental). As seen in our LC-ESI-MS crude extract profiles (Figure 4) the purple variety contained large amounts of violet betacyanins such as 15*S*/15*R*-betanin ($[M+H]^+$: *m/z* 551), but 15*S*/15*R*-gomphrenin was not detectable. For the purple variety, neo-betanin (*m/z*

549) was seen and is known as a classical betanin degradation product (*1*, *26*). Indicaxanthin (prolin-bx) as chemotaxonomic marker pigment ($[M+H]^+$: *m/z* 309) was present in all crude pigment extracts of the three color varieties. Here, the analysis of the orange crude extract detected a low amount of 15*S*/15*R*-betanin, but for the yellow variety it was not detected by selective ion monitoring (Figure 4). Minor concentrated betaxanthin pigments such as phenylalanine-bx (*m/z* 359), and γ-amino-butyric acid-bx (*m/z* 297) were detected in all extracts. Resuming the pigment profiles, yellow and orange *Opuntia* extracts are less complex and more suitable for the preparative recovery of betaxanthins. The LC-ESI-MS/MS analysis of a reference crude extract of *Opuntia ayrampo* (syn. *Tunilla soehrensii*) detected a rather simple and clean pigment profile consisting of 15*S*/15*R*-betanin, and 15*S*/15*R*-phyllocactin (malonyl-betanin) with $[M+H]^+$ at *m/z* 637 (Figure 4). All *Opuntia* extracts contained lipophilic compounds with ion-signals of even numbered pseudomolecular ions suggesting an uneven content of nitrogen atoms in the molecules ($[M+H]^+$: *m/z* 484, *m/z* 498, *m/z* 406, *m/z* 288). The discussion of some MS/MS data of suggested alkaloid structures will be presented with Figure 10.

IP-HSCCC Chromatography of C18 Reversed Phase Pre-Cleaned Pigment Extracts of *Opuntia ficus*

The generated pigment extracts of the violet, orange, and yellow varieties were separated by means of IP-HSCCC with the solvent system TBMe – n-butanol – acetonitrile – water (0.7% HFBA) (Figure 1). Injection amounts were in the range between 850 and1020 mg.

Fractionation of 15R/15S-Betanin, 15R/15S-Gomphrenin, Indicaxanthin, and Other Compounds from Violet *Opuntia ficus* Fruits

In the IP-HSCCC chromatography betacyanins were eluted rapidly (190-230 min). Glycosidation of these components cause a more polar character and ion-pair effects were less active than for betaxanthins. Indicaxanthin was clearly detected at λ 440 nm (320-360 min) (Figure 5A-5C). For the recovery of the whole 'polarity window' of the injected sample, the *extrusion mode* was started after 500 min experimental time. The IP-HSCCC fractions of *Opuntia* extracts (yields cf. Experimental) were investigated by LC-ESI-MS/MS using HPLC with a C18 encapped reversed phase. The HSCCC chromatography had focussed minor concentrated compounds to specific fractions, therefore the metabolite detection limits for LC-ESI-MS/MS were more sensitive than for the analysis of the respective crude extracts. Therefore, two different chromatographic dimensions (IP-HSCCC - C18 HPLC-ESI-MS) enabled the detection of many unknown trace compounds – structurally not related to the pigment structures. Violet color hues were visible in the tubes of the fraction collector. Traces of15*S*/15*R*-betanin were confirmed by LC-ESI-MS in orange and also in yellow *Opuntia* extract.

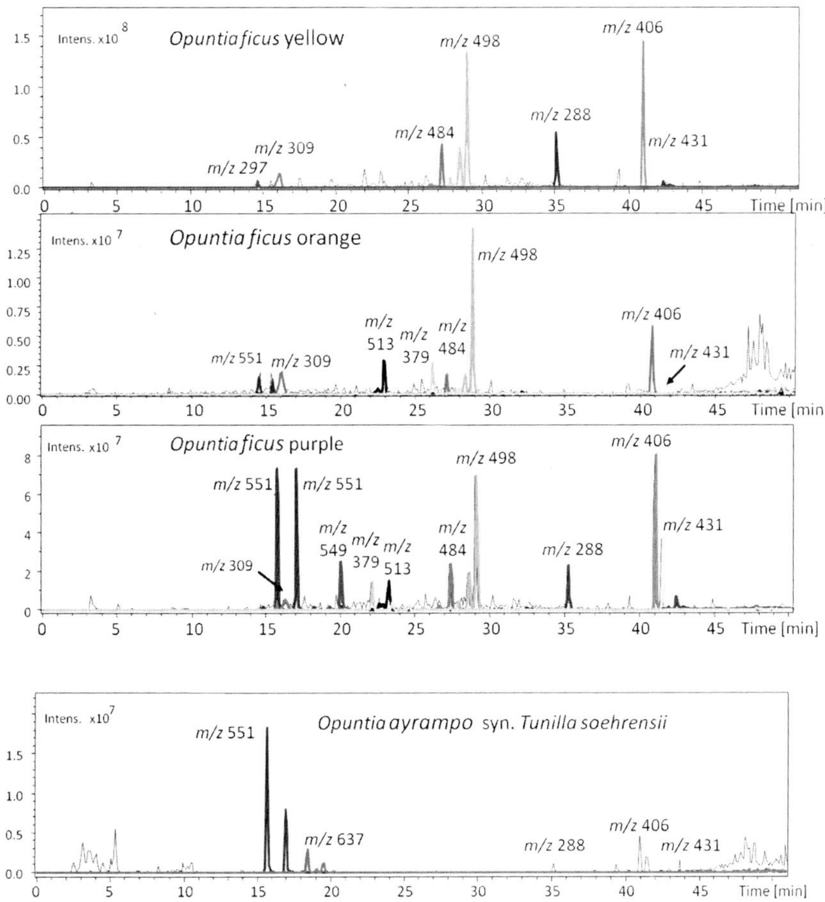

Figure 4. LC-ESI-MS pigment profiles of Opuntia spp. extracts measured in positive ionization mode (m/z 50-2200).

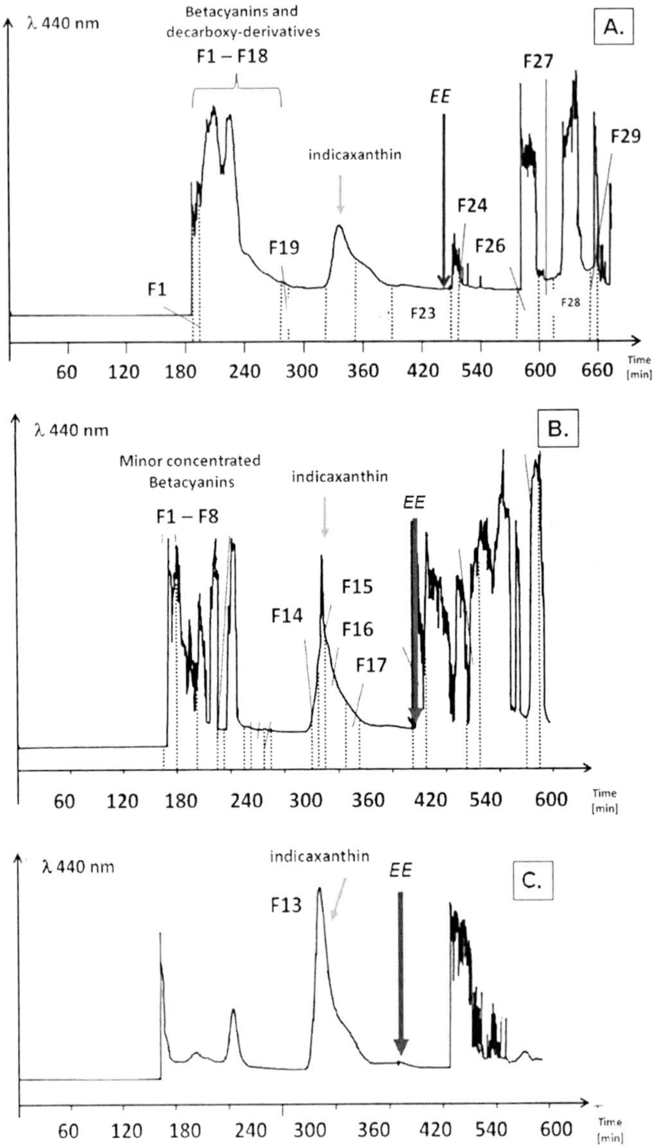

Figure 5. C18 MPLC pre-cleaned Opuntia pigment extracts injected to IP-HSCCC. Solvent system TBMe – n-butanol – acetonitrile – water 1:3:1:5 (0.7% HFBA). Elution – extrusion mode, detection λ 440 nm. A. purple Opuntia (1020 mg), B. yellow Opuntia (850 mg), C. orange Opuntia (980 mg). EE = Elution – Extrusion.

Off-Line LC-ESI-MS/MS of IP-HSCCC Fractions (Purple *Opuntia* Extract)

The polar IP-HSCCC fractions (F1-F4) investigated by LC-ESI-MS/MS analysis detected a complex metabolite profile (Figure 6). In fraction F4, the most polar pigment glucosyl-betanin was detected with $[M+H]^+$ at *m/z* 713 with its two epimeric forms (15*S*/15*R*). MS/MS data clearly revealed the cleavage of a hexose unit with a neutral loss Δ*m/z* 162 to *m/z* 551. This compound was tentatively identified as betanidin 5-*O*-*β*-sophoroside, detected previously in *Hylocereus polyrizhus* fruits (*7*). All other detected compounds remained unknown and seemed not to be related to betalain structures. Fraction F5 contained the betanin degradation product neo-betanin (*m/z* 549) with an aromatized pyridine system (*1*, *26*). The fractions F9 to F12 already appeared to be very clean in the contents of 15*S*/15*R*-betanin (yield: ~ 85 mg) (Figure 6). Observing the respective selected ion trace at *m/z* 551 (Figure 7), it was recognized that IP-HSCCC partly fractionated 15*S*- from 15*R*-betanin. LC-ESI-MS detection showed that the epimer 15*S*-betanin had a stronger affinity to the liquid stationary HSCCC phase and eluted with higher concentration in fraction F12 and a partly epimeric fractionation was suggested. The sequence of C18-HPLC elution of the epimers 15*S* and15*R*-betanin seemed to be opposite to IP-HSCCC, and both separation techniques could be used in a complimentary way. F9 contained a polar compound with $[M+H]^+$ at *m/z* 358 and MS/MS with Δ*m/z* 162 to *m/z* 196 which was recognized as *cyclo*-dopa-5-O-glucoside as a precursor of betacyanin biosynthesis (*27*) (Figure 8).

The selected ion trace for *m/z* 551 (Figure 7) in F13 detected four clearly separated peaks, and suggested the HSCCC co-elution of 15S/15R-betanin with 15S/15R-gomphrenin (glc-6-O-betanidin). F14 solely contained the gomphrenin epimers. F13 contained more pigments such as 15*S*/15*R*-phyllocactin (malonyl-betanin), with *m/z* 637 and neutral loss of Δ*m/z* 86 in the MS/MS mode. These epimers are typical pigments of cacti fruits e.g. *Hylocereus polyrhizus* (*7*, *18*). Three different mono-decarboxy-betanins were detected at *m/z* 507 related to17-, 15-, and 2-mono-decarboxy-betanin and are degradation products caused during the pigment extraction and handling for HSCCC. The pyridine-derivative with $[M+H]^+$ at *m/z* 459 was already known from processed red beet extracts (*1*) and was clearly indentified by the MS/MS data with *m/z* 297 for the residual aglycone (Figure 8). The most frequent decarboxylation positions in betanin are at C-2 and C-17, therefore it can be stated that this is 2,17-bidecarboxy-2,3-dehydro-neobetanin (*28*). Fraction F14 contained eucomic acid with the selected ion-trace at *m/z* 241, a polar phenolic organic acid, typically occurring in green *Opuntia* stem and cladode materials (*29*, *30*), however piscidic acid (*29–31*) was not detected in the HSCCC-fractions. The LC-ESI-MS analysis of F20 displayed a high compound complexity and the selected ion trace revealed at *m/z* 415 a lipophil break-down product of betanin known from red beet processings. MS/MS confirmed the cleavage of a glucose unit to the fragment-ion at *m/z* 253 (Figure 8) (*1*). Fractions F21 and F22 contained indicaxanthin with $[M+H]^+$ at *m/z* 309 which appeared to contain slightly impurities as detected by the LC-ESI-MS monitoring. F22 contained γ-aminobutyric acid-bx $[M+H]^+$ at *m/z* 297. F25 contained the more lipophilic betacyanin aglycone 15*S*/15*R*-betanidin (*m/z* 389), and also phenylalanine-bx (*m/z* 359), typically known for *Opuntia* fruits (*2*, *3*, *25*). Betanidin could also be seen as an enzymatic cleavage product generated during juice processing and residual β-glycosidase activity. F26 contained two epimers of a lipophilic betacyanin with $[M+H]^+$ at *m/z* 727 and neutral loss of $\Delta m/z$ 176 indicated the cleavage of one feruloyl-substitution. It was not possible to clarify if these compounds were feruloyl-betanin (lampranthin II) or gomphrenin III. A very typical pigment for cacti fruits is also 15*S*/15*R*-hylocerenin (hydroxyl-methyl-glutaryl-betanin) with $[M+H]^+$ at *m/z* 695. This ion-trace was absent in all investigated IP-HSCCC fractions of the investigated *Opuntia* extracts (*4*, *18*).

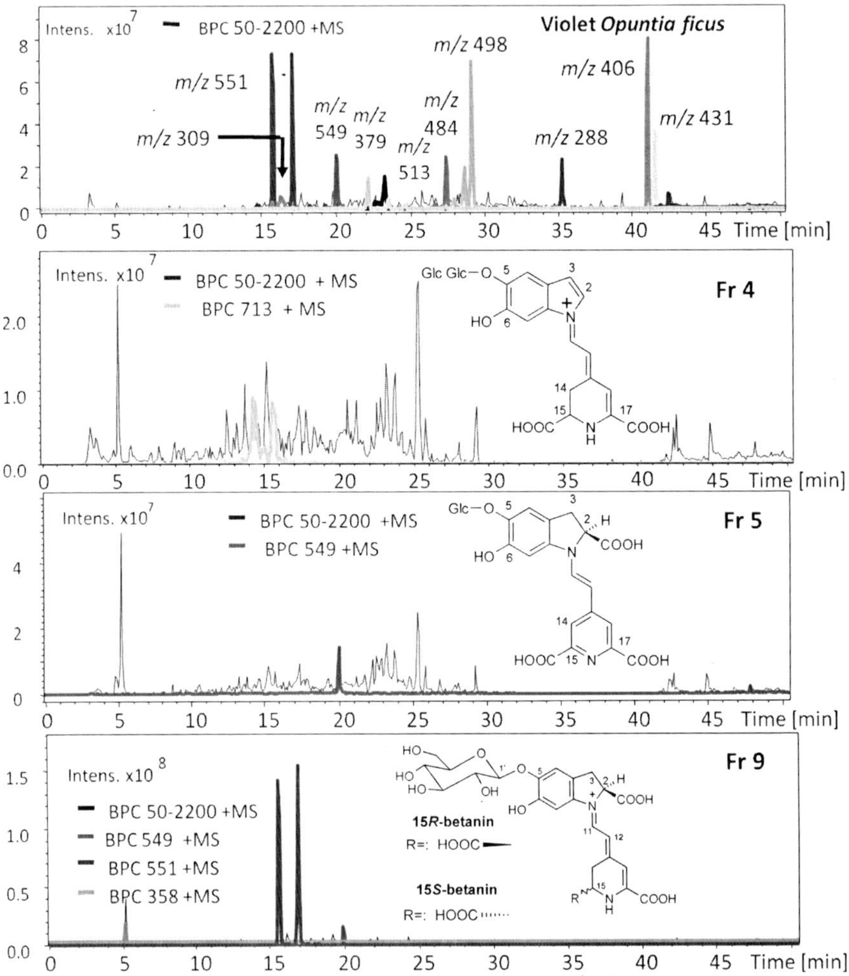

Figure 6. Off-line LC-ESI-MS pigment profiles of recovered IP-HSCCC fractions from fortified purple Opuntia fruit pigment extract.

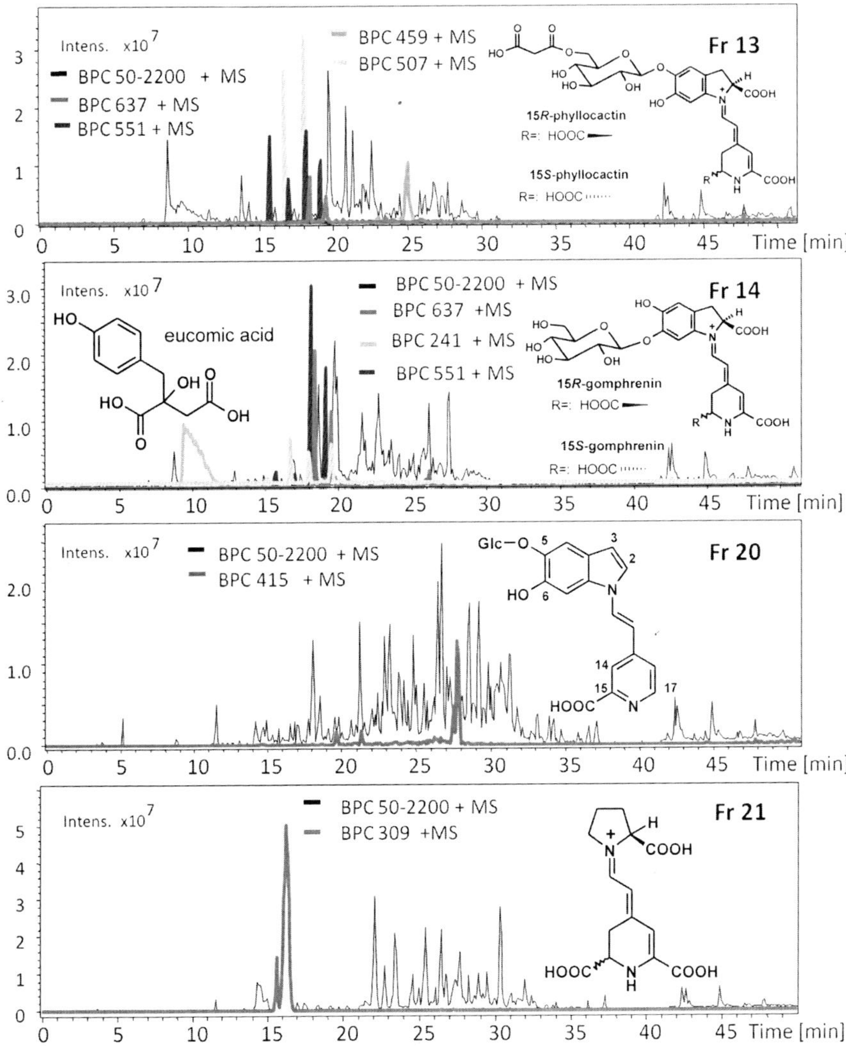

Figure 6. (continued) Off-line LC-ESI-MS pigment profiles.

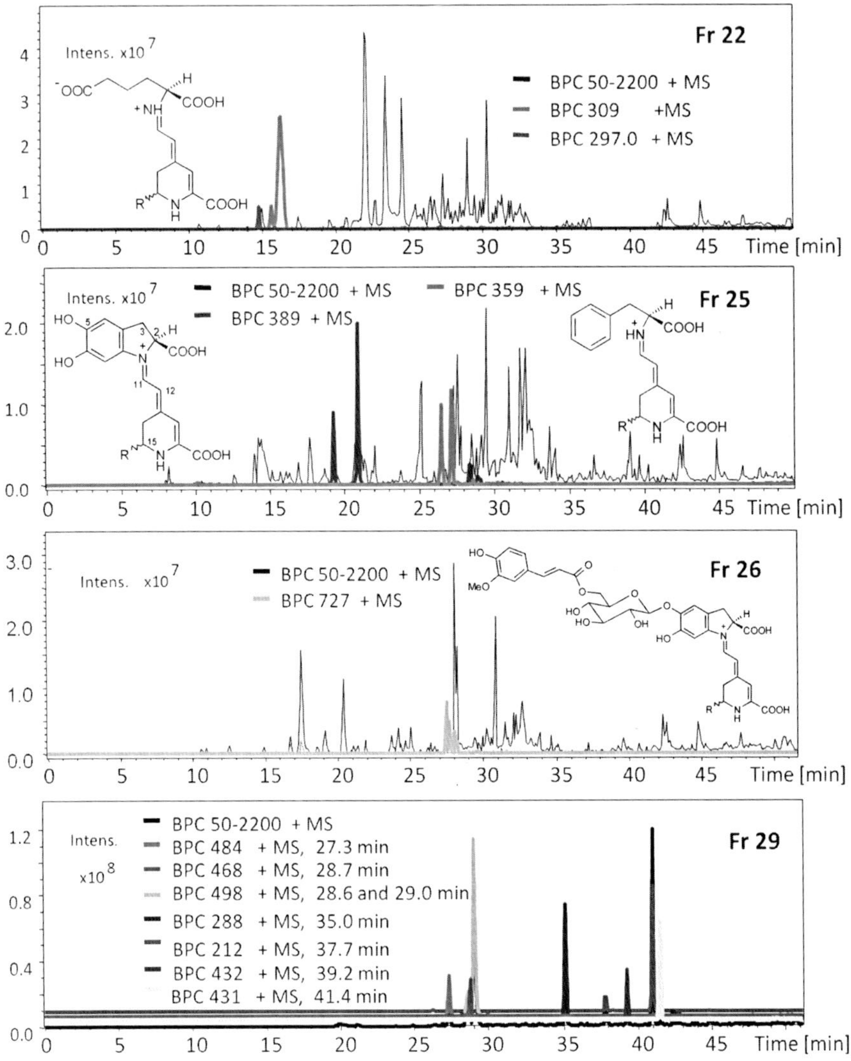

Figure 6. (continued) Off-line LC-ESI-MS pigment profiles.

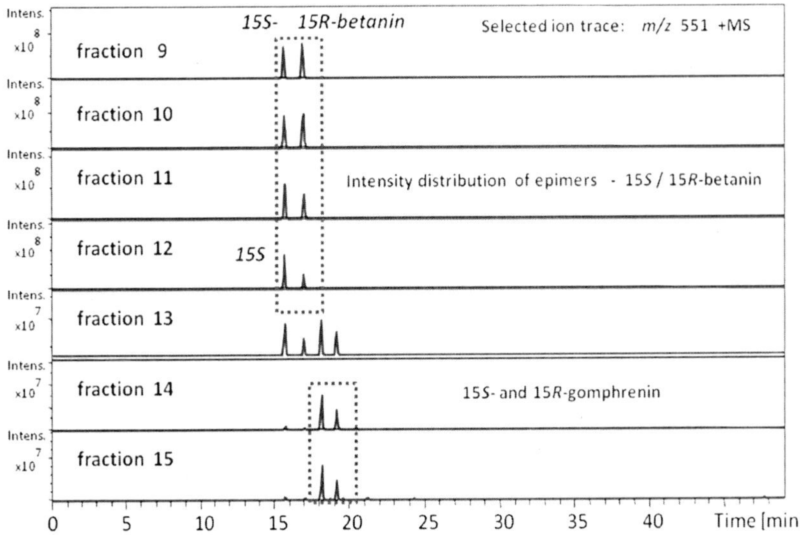

Figure 7. LC-ESI-MS (positive mode) of the IP-HSCCC fractions F9-F15 (purple Opuntia C18 MPLC pre-cleaned extract) with the selected ion trace at m/z 551 detected the fractionation of epimers 15S- and 15R-betanin, and the separation of the positional isomer 15S/15R-gomphrenin.

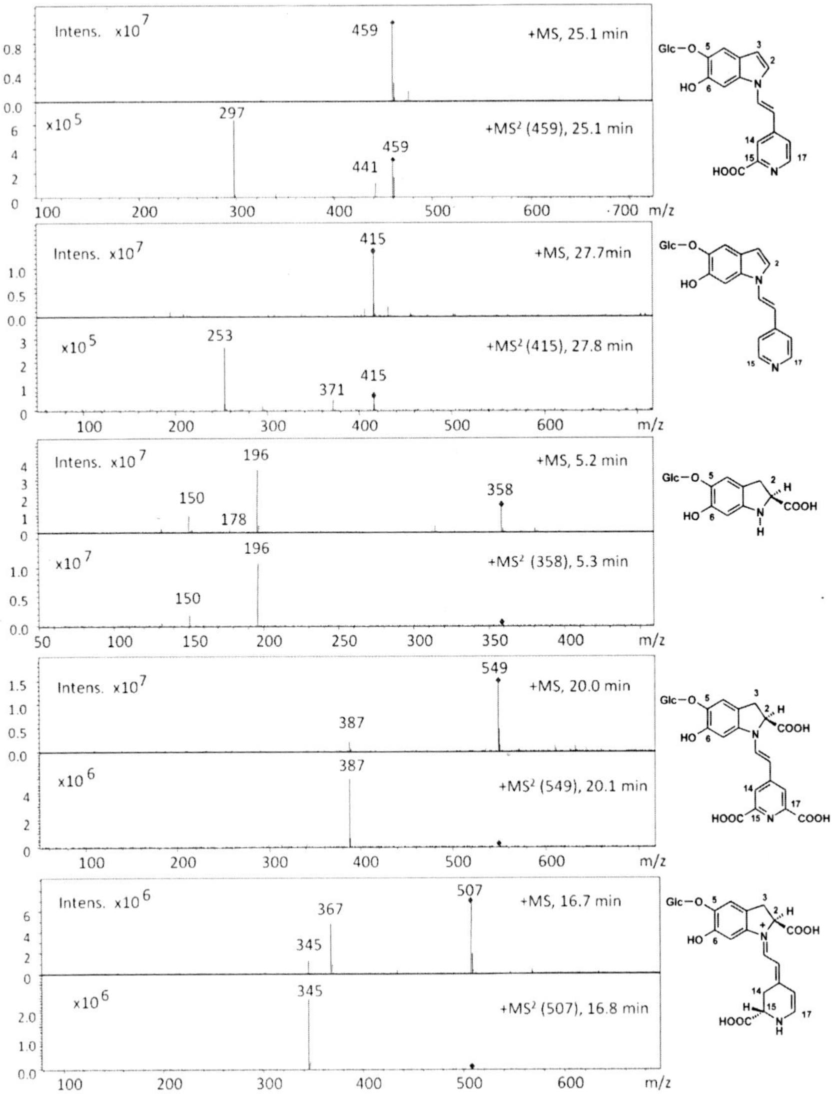

Figure 8. LC-ESI-MS/MS spectra (pos. mode) of alterated betacyanin pigments from purple Opuntia ficus.

The last fractions F28 and F29 were highly concentrated in compounds already detected in the LC-ESI-MS runs of the *Opuntia* crude extracts and were potentially related to alkaloids (cf. MS/MS data Figure 10).

Indicaxanthin Purity from IP-HSCCC Fractions

The analysis of the IP-HSCCC fractions containing high amounts of indicaxanthin were evaluated by LC-ESI-MS. The highest purities were achieved from the orange and yellow *Opuntia ficus* C18 pre-cleaned pigment extracts and yields were between 100-130 mg of indicaxanthin for a single CCC run (indicaxantin in purple *Opuntia* cf. Figure 6, orange and yellow *Opuntia* cf. Figure 9).

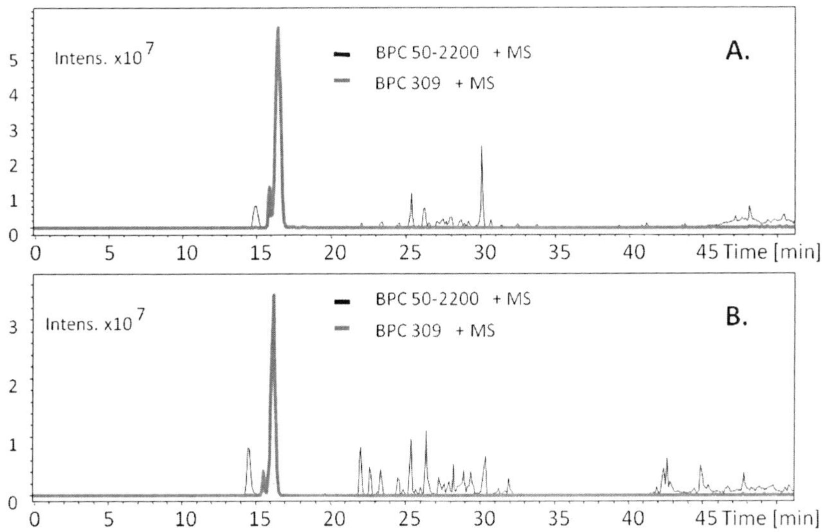

Figure 9. LC-ESI-MS analysis (pos. mode) for purity evaluation of IP-HSCCC fractions containing highest amounts of indicaxanthin. A: indicaxanthin in F13 (orange Opuntia); B: indicaxanthin in F15 (yellow Opuntia) (5).

Unidentified Compounds in Late Eluting IP-HSCCC Fractions

The final eluting fractions of the IP-HSCCC runs (e.g. F28 and F29: purple *Opuntia,* Figure 6) contained lipophilic compounds already detected by LC-ESI-MS/MS in all crude pigment extracts of yellow, orange and purple *Opuntia* fruits (Figure 4). The selected ion traces with the even numbered pseudomolecular ion signals ($[M+H]^+$) in the positive ESI-MS mode suggested that these compounds contain uneven numbers of nitrogen.

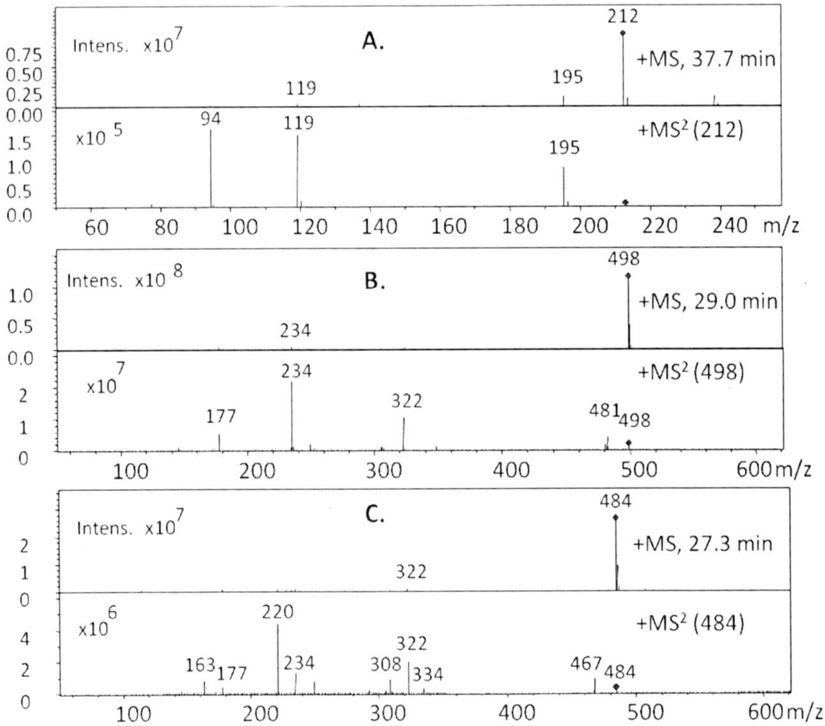

Figure 10. LC-ESI-MS/MS spectra (pos. mode) of some proposed alkaloids in Opuntia ficus. A. MS/MS: mescaline; B: MS/MS: berberine-O-glucuronide; C: MS/MS berberine-O-glucoside.

The small peak at 37.7 min in the LC-ESI-MS/MS chromatogram of HSCCC fraction F29 (Figure 6) with $[M+H]^+$ at *m/z* 212, and the MS/MS fragment ion at *m/z* 195 ($[M+H-NH_3)$) indicated the occurrence of the alkaloid mescaline - a *β*-phenethylamine - already known from various *Opuntia* species (*32–34*). The abundant peak at Rt 29.0 min with $[M+H]^+$ at *m/z* 498 and MS/MS at *m/z* 322 was in good accordance to published ESI-MS/MS data of berberine-*O*-glucuronide (cf. Figure 10) (*35*, *36*). Signal *m/z* 322 could be the aglycone-moiety berberrubine, and the neutral loss cleavage of Δ*m/z* 176 indicated a substitution with glucuronic acid. A lower concentrated positional isomer was detected at Rt 28.5 min with identical MS and MS/MS-data. For the compound at Rt 27.3 min with $[M+H]^+$ at *m/z* 484 and fragment ion at *m/z* 322 (neutral loss Δ*m/z* 162) berberine-*O*-glucoside was proposed. Additionally detected ion signals at *m/z* 288, *m/z* 468, *m/z* 406, *m/z* 432, and *m/z* 431 could be structurally related to alkaloids as well (Figure 6). However, all these LC-ESI-MS/MS detected compounds will require preparative isolation, and 1D/2D-NMR spectroscopy for a complete structural characterization.

Conclusions

The two dimensional chromatography approach (IP-HSCCC - C18-HPLC) had demonstrated that fruits from *Opuntia ficus* - independently from the color variety - contain a very large number of structurally still unknown and potentially bioactive metabolites partly related to betalain pigments. Interestingly, phytochemical research with modern spectroscopical methods on very minor constituents in cacti fruit, such as *Opuntia* is still missing. The detected metabolite profiles of non-pigment related structures will require intensive preparative clean-up procedures and complete structure eludication by modern 1D/2D-NMR techniques.

From point of stability it was clearly seen that genuine pigments easily undergo changes in their constitution and that also gentle processing steps could not prevent the induction of various betalain degradation products. The presence of the dehydrogenated derivatives of betanin (neo-betanin, 2,17-bi-decarboxy-2,3-dehydro-neobetanin and 2,15,17-tri-decarboxy-2,3-dehydro-neobetanin) in the collected fractions from the HSCCC separations of *Opuntia* extracts suggested that the applied solvent systems are still too reactive for betanin and a decarboxylative oxidation of the pigments took place. Therefore, we are still looking for a less degrading solvent system for betalain separations. The presence of perfluorinated carboxylic acids in the applied solvents is necessary to achieve the best pigment separation. Finding a solvent system not containing these toxic acids will also make the application of betalains for medicinal tests or food science at much higher extent possible.

Aim of this study was to demonstrate that IP-HSCCC could isolate principal pigments from *Opuntia* in the range of 100 mg, which then can be used as authentic reference compounds to monitor pigment changes during technological processings of *Opuntia* fruit products (*37*).

Industrial scale *countercurrent-separations* or *-extractions* are feasible and could deliver pigments for nutraceutical uses, and also for substitution of synthetic dyes in foods (*38*).

In the future, a promising approach might be to test purified pigments in a broad portefolio of biological assays to screen for specific biofunctionalities.

Especially *Opuntia* fruits are underestimated in their value for human nutrition. Fruits of *Opuntia* are relatively easy accessible in sub-tropical and tropical areas (*29*, *37*, *39–42*). Large areas are covered with *Opuntia* spp. and plantation is not necessarily needed for cultivation. Wild collections delivered larger amounts of fruits for medium size production scale, such as accomplished in Mexico (*43–45*). Associations of farmer communities have started small production sides and designed new *Opuntia* fruit products.

Cacti plants could be promoted to be a more important food crop in the future. An increased awareness about the benefits of cacti fruit consumption is needed to stimulate the market and development of attractive product lines such as juices, jams, smoothies, natural colorants for foods, functional foods, supplements with certain health benefits could be done (*42*, *45–47*).

Cactus fruit production could be seen as a small answer to help guarantee sustainability in the food chains of specific arid regions suffering from desertification like in North-America, Latin America, Southern Europe, Africa, and also China.

Acknowledgments

The authors G.J. and S.V.N. are grateful for the research grant provided by the DAAD-CONACyT research foundations (PROALMEX D/07/10252 PPP).

The research was also supported by the European Union through the European Social Fund within "Cracow University of Technology development program - top quality teaching for the prospective Polish engineers; University of the 21st century" project (contract no.UDA-POKL.04.01.01-00-029/10-00).

References

1. Nemzer, B.; Pietrzkowski, Z.; Spórna, A.; Stalica, P.; Thresher, W.; Michalowski, T.; Wybraniec, S. Betalainic and nutritional profiles of pigment-enriched red beet root (*Beta vulgaris* L.) dried extracts. *Food Chem.* **2011**, *127*, 42–53.
2. Stintzing, F. C.; Schieber, A.; Carle, R. Identification of betalains from yellow beet (*Beta vulgaris* L.) and cactus pear [*Opuntia ficus indica* (L.) Mill.] by high-performance liquid chromatography-electrospray ionization mass spectrometry. *J. Agric. Food Chem.* **2002**, *50*, 2302–2307.
3. Stintzing, F. C.; Herbach, K. M.; Mosshammer, M. R.; Carle, R.; Yi, W.; Sellappan, S.; Akoh, C. C.; Bunch, R.; Felker, P. Color, betalain pattern, and antioxidant properties of cactus pear (*Opuntia* spp.) clones. *J. Agric. Food Chem.* **2005**, *53*, 442–451.
4. Kugler, F.; Stintzing, F. C.; Carle, R. Identification of betalains from petioles of differently colored swiss chard (*Beta vulgaris* L. ssp. cicla [L.] Alef. Cv. bright lights) by high-performance liquid chromatography-electrospray ionization mass spectrometry. *J. Agric. Food Chem.* **2004**, *52*, 2975–2981.
5. Cai, Y.; Sun, M.; Corke, H. Identification and distribution of simple and acylated betacyanins in the Amaranthaceae. *J. Agric. Food Chem.* **2001**, *49*, 1971–1978.
6. Svenson, J.; Smallfield, B. M.; Joyce, N. I.; Sansom, C. E.; Perry, N. B. Betalains in red and yellow varieties of the Andean tuber crop Ulluco (*Ullucus tuberosus*). *J. Agric. Food Chem.* **2008**, *56*, 7730–7737.
7. Wybraniec, S.; Platzner, I.; Geresh, S.; Gottlieb, H. E.; Haimberg, M.; Mogilnitzki, M.; Mizrahi, Y. Betacyanins from vine cactus *Hylocereus polyrhizus*. *Phytochemistry* **2001**, *58*, 1209–1212.
8. Guzmán-Maldonado, S. H.; Herrera-Hernández, G.; Hernández-López, D.; Reynoso-Camancho, R.; Guzmán-Tovar, A.; Vaillant, F.; Brat, P. Physicochemical, nutritional and functional characteristics of two underutilised fruit cactus species (*Myrtillocactus*) produced in central Mexico. *Food Chem.* **2010**, *121*, 381–386.

9. Barrera, F. A. G.; Reynoso, C. R.; González de Mejía, E. Stability of betalains extracted from garambullo (*Myrtillocactus geometrizans*). *Food Sci. Technol. Int.* **1998**, *4*, 115–120.
10. Strack, D.; Vogt, T.; Schliemann, W. Recent advances in betalain research. *Phytochemistry* **2003**, *62*, 247–269.
11. Butera, D.; Tesoriere, L.; Di Gaudio, F.; Bongiorno, A.; Allegra, M.; Pintaudi, A. M.; Kohen, R.; Livrea, M. Antioxidant activities of sicilian prickly pear (*Opuntia ficus indica*) fruit extracts and reducing properties of its betalains: Betanin and indicaxanthin. *J. Agric. Food Chem.* **2002**, *50*, 6895–6901.
12. Tesoriere, L.; Fazzari, M.; Angileri, F.; Gentile, C.; Livrea, M. In-vitro digestion of betalainic foods. Stability and bioaccessibility of betaxanthins and betacyanins and antioxidative potential of food digesta. *J. Agric. Food Chem.* **2008**, *56*, 10487–10492.
13. Tesoriere, L.; Butera, D.; Allegra, M.; Fazzari, M.; Livrea, M. Distribution of betalain pigments in red blood cells after consumption of cactus pear fruits and increased resistance of the cells to ex vivo induced oxidative hemolysis in humans. *J. Agric. Food Chem.* **2005**, *53*, 1266–1270.
14. Galati, E. M.; Mondello, M. R.; Giuffrida, D.; Dugo, G.; Micelli, N.; Pergolizzi, S.; Taviano, M. F. Chemical characterization and biological effects of Sicilian *Opuntia ficus indica* (L.) Mill. fruit juice: Antioxidant and antiulcerogenic activity. *J. Agric. Food Chem.* **2003**, *51*, 4903–4908.
15. Wickert, T. Ph.D. Thesis, University Würzburg, Würzburg, Germany, 2006.
16. Tesoriere, L.; Fazzari, M.; Allegra, M.; Livrea, M. A. Biothiols, taurine, and lipid-soluble antioxidants in the edible pulp of Sicilian cactus pear (*Opuntia ficus indica)* fruits and changes of bioactive juice components upon industrial processing. *J. Agric. Food Chem.* **2005**, *53*, 7851–7856.
17. Jerz, G.; Skotzki, T.; Fiege, K.; Winterhalter, P.; Wybraniec, S. Separation of betacyanins and betaxanthins from berries of *Phytolacca americana* by ion-pair high-speed counter-current chromatography. *J. Chromatogr., A* **2008**, *1190*, 63–73.
18. Wybraniec, S.; Stalica, P.; Jerz, G.; Klose, B.; Gebers, N.; Winterhalter, P.; Spórna, A.; Szaleniec, M.; Mizrahi, Y. Separation of polar betalain pigments from cacti fruits of *Hylocereus polyrhizus* by ion-pair high-speed counter-current chromatography. *J. Chromatogr., A* **2009**, *1216*, 6890–6899.
19. Wybraniec, S.; Jerz, G.; Gebers, N.; Winterhalter, P. Ion-pair high speed countercurrent chromatography fractionation of a high-molecular weight variation of acyl-oligosaccharide linked betacyanins from purple bracts of *Bougainvillea glabra*. *J. Chromatogr., B* **2010**, *878*, 538–550.
20. Jerz, G.; Wybraniec, S.; Gebers, N.; Winterhalter, P. Target guided separation of *Bougainvillea glabra* betacyanins by direct coupling of preparative ion-pair high-speed countercurrent chromatography and electrospray ionization mass-spectrometry. *J. Chromatogr., A* **2010**, *1217*, 4544–4554.
21. Ito, Y.; Conway, W. D., Eds.; *High-Speed Countercurrent Chromatography*; Chemical Analysis 132; Wiley: New York, 1996.

22. Ito, Y. Golden rules and pitfalls in selecting optimum conditions for high-speed counter-current chromatography. *J. Chromatogr., A* **2005**, *1065*, 145–168.
23. Berthod, A.; Ruiz-Angel, M. J.; Carda-Broch, S. Elution-extrusion countercurrent chromatography. Use of the liquid nature of the stationary phase to extend the hydrophobicity window. *Anal. Chem.* **2003**, *75*, 5886–5894.
24. de Beer, D.; Jerz, G.; Joubert, E.; Wray, V.; Winterhalter, P. Isolation of isomangiferin from honeybush (*Cyclopia subternata*) using high-speed counter-current chromatography and high-performance liquid chromatography. *J. Chromatogr., A* **2009**, *1216*, 4282–4289.
25. Castellanos-Santiago, E.; Yahia, E. M. Identification and quantification of betalains from the fruits of 10 mexican prickly pear cultivars by high-performance liquid chromatography and electrospray ionization mass spectrometry. *J. Agric. Food Chem.* **2008**, *56*, 5758–5764.
26. Alard, D.; Wray, V.; Grotjahn, L.; Reznik, H.; Strack, D. Neobetanin: Isolation and identification from *Beta vulgaris*. *Phytochemistry* **1985**, *24*, 2383–2385.
27. Herbach, K. M.; Stintzing, F. C.; Carle, R. Identification of heat-induced degradation products from purified betanin, phyllocactin and hylocerenin by high-performance liquid chromatography/electrospray ionization mass spectrometry. *Rapid Commun. Mass Spectrom.* **2005**, *19*, 2603–2616 (Erratum: **2006**, *20*, 1822).
28. Wybraniec, S.; Michałowski, T. New pathways of betanidin and betanin enzymatic oxidation. *J. Agric. Food Chem.* **2011**, *59*, 9612–9623.
29. Stintzing, F. C.; Carle, R. Cactus stems (*Opuntia* spp.): A review on their chemistry, technology, and uses. *Mol. Nutr. Food Res.* **2005**, *49*, 175–194.
30. Heller, W.; Tamm, C. 189. Isolierung, konstitution und synthese der (R) - (—)-eucominsäure. *Helv. Chim. Acta* **1974**, *57*, 1766–1784.
31. Heller, W.; Tamm, C. Fukiinsäure und 3′-*O*-methyl-fukiinsäure, zwei phenolische hydroxycarbonsäuren aus *Piscidia Erythrina*. *Helv. Chim. Acta* **1975**, *58*, 974–979.
32. Meyer, B. N.; Mohamed, Y. A. H.; McLaughlin, J. L. *β*-Phenethylamines from the cactus genus *Opuntia*. *Phytochemistry* **1980**, *19*, 719–720.
33. Beyer, J.; Peters, F. T.; Kraemer, T.; Maurer, H. Detection and validated quantification of nine herbal phenalkylamines and methcathinone in human blood plasma by LC-MS/MS with electrospray ionization. *J. Mass Spectrom.* **2007**, *42*, 150–160.
34. Ma, W. W.; Jiang, X. Y.; Cooks, R. G.; McLaughlin, J. L.; Gibson, A. C.; Zeylemaker, F.; Ostolaza, C. N. Cactus alkaloids, LXI. Identification of mescaline and related compounds in eight additional species using TLC and MS/MS. *J. Nat. Prod.* **1986**, *49*, 735–737.
35. Zuo, F.; Nakamura, N.; Akao, T.; Hattori, M. Pharmacokinetics of berberine and its main metabolites in conventional and pseudo germ-free rats determined by liquid chromatography/ion trap mass spectrometry. *Drug Metab. Dispos.* **2006**, *34*, 2064–2072.

36. Qiu, F.; Zhu, Z.; Kang, N.; Piao, S.; Qin, G.; Yao, X. Isolation and indentification of urinary metabolites of berberine in rats and humans. *Drug Metab. Dispos.* **2008**, *36*, 2159–2165.
37. Stintzing, F. C.; Carle, R. Betalains – Emerging prospects for food scientists. *Trends Food Sci. Technol.* **2007**, *18*, 514–525.
38. Sutherland, I.; Hewitson, P.; Ignatova, S. New 18-l process-scale counter-current chromatography centrifuge. *J. Chromatogr., A* **2009**, *1216*, 4201–4205.
39. Fuentes-Rodiguez, J.; Murillo-Soto, M. Prickly Cactus 2010: A prospective view. *J. Prof. Assoc. Cactus Dev.(PACD)* **1996**, *1*, 67–72.
40. Pimienta-Barrios, E. Prickly pear (*Opuntia* spp.): A valuable fruit crop for semi-arid lands of Mexico. *J. Arid Environ.* **1994**, *28*, 1–11.
41. Nazareno, M. A.; Padrón Pereira, C. A. Nuevas tecnologías desarrolladas para el aprovechamiento de las cactáceas en la elaboración de alimentos. Componentes funcionales y propiedades antioxidants. *Rev. Venez. Cienc. Tecnol. Aliment.* **2011**, *2*, 202–238.
42. Stintzing, F. C.; Schieber, A.; Carle, R. Phytochemical and nutritional significance of cactus pear. *Eur. Food Res. Technol.* **2001**, *212*, 396–407.
43. Griffith, M. P. The origins of an important cactus crop, *Opuntia ficus-indica* (Cactaceae): New molecular evidence. *Am. J. Bot.* **2004**, *91*, 1915–1921.
44. Russel, R. E.; Felker, P. The prickly-pears (*Opuntia* spp. Cactaceae): A source of human and animal food in semiarid regions. *Econ. Bot.* **1987**, *41*, 433–445.
45. Casas, A.; Cruse-Sanders, J.; Morales, E.; Otero-Arnaiz, A.; Valiente-Banuet, A. Maintenance of phenotypic and genotypic diversity in managed populations of *Stenocereus stellatus* (Cactaceae) by indigenous peoples in Central Mexico. *Biodivers. Conserv.* **2006**, *15*, 879–898.
46. Moreno, D. A.; García-Viguera, C.; Gil, J. I.; Gil-Izquierdo, A. Betalains in the era of global agri-food science technology and nutritional health. *Phytochem. Rev.* **2008**, *7*, 261–280.
47. Obón, J. M.; Castellar, M. R.; Alacid, M.; Fernández-López, J. A. Production of a red-purple food colorant from *Opuntia stricta* fruits by spray-dyring and its application in food model systems. *J. Food Eng.* **2009**, *90*, 471–479.

Chapter 2

High Hydrostatic Pressure Processing as a Strategy To Increase Carotenoid Contents of Tropical Fruits

Carmen Hernández-Brenes,*[,1] Perla A. Ramos-Parra,[1] Daniel A. Jacobo-Velázquez,[1] Raul Villarreal-Lara,[1] and Rocío I. Díaz-De la Garza[1,2]

[1]Department of Biotechnology and Food Engineering, School of Biotechnology and Food, Tecnológico de Monterrey-Campus Monterrey, E. Garza Sada 2501 Sur, C.P. Monterrey, N.L. 64849, México
[2]Centro de Agrobiotecnología, Departamento de Agrobiotecnología y Agronegocios, Tecnológico de Monterrey-Campus Monterrey, Av. Eugenio Garza Sada 2501 Sur, Monterrey, N.L. 64849, México
***E-mail: chbrenes@itesm.mx. Tel.: +52-818-328-4262, Ext. 109. Fax: +52-818-328-4136.**

Avocado (*Persea americana*), papaya (*Carica papaya* cv. Maradol) and mango (*Mangifera indica L.*) contain high levels of carotenoid molecules that protect human cells against the detrimental effects of reactive oxygen species, among other bioactive compounds. High hydrostatic pressure (HHP) processing is an effective non-thermal alternative to stabilize tropical fruit pulps due to retention of flavor and desirable sensory characteristics. Moreover, certain HHP treated fruits and vegetables products show higher concentrations of extractable carotenoids than their corresponding fresh product unprocessed samples. Chapter objectives were to evaluate the HHP processing stability of carotenoids from avocado, papaya and mango pulps. HHP processing (600 MPa/3 min) caused an increase in total carotenoid concentrations (~56%) for avocado pulp. Papaya pulps also contained higher levels of precursors (215%), xanthophylls (219%) and carotenes (164%) while mango carotenoids remained unchanged. Higher

© 2013 American Chemical Society

concentrations of extractable carotenoids have been attributed to possible changes in the permeability of cells and chloroplast membranes induced by HHP processing. Likewise, it has been reported that HHP application causes oxidative stress within plant tissues, and RNA molecules remain intact in pressurized samples suggesting that the cells are metabolically active and biosynthesizing plant cell antioxidants.

Introduction

High hydrostatic pressure (HHP) processing is an effective alternative to stabilize fruit pulps due to retention of flavor and desirable sensory characteristics. Likewise, it has been reported that the concentration of vitamin and pigments in HHP processed fruits is similar to those found in fresh (non-processed) samples. The retention of bioactive molecules is mainly attributed to the low effect that HHP has on low molecular weight compounds (*1*, *2*). In addition, reports have shown that pressurization treatment increased the concentrations of extractable carotenoids (*3*, *7*, *10*, *37*) and it has been observed that carotenoids present in HHP processed products are more bioavailable than those present in unprocessed samples (*36*).

Sánchez-Moreno *et al.* (*3*) evaluated the effect of different HHP and temperature treatments (100 MPa/60°C/5 min; 350 MPa/30°C/2.5 min; 400 MPa/40°C/1 min) on the extractability of orange juice carotenoids with pro-vitamin A activity (β- and α- carotene; β- and α-cryptoxanthin) and xanthophylls (zeaxanthin and lutein). The authors reported that as the pressure treatment intensity increased, the quantity of carotenoids extracted was higher, obtaining ~32% higher concentration of total carotenoids in the orange juice processed at 400 MPa/40°C/1 min as compared with the non-processed samples (*3*). Likewise, De Ancos *et al.* (*4*) evaluated the effects of HHP treatments (50-400 MPa/25°C/15 min) on the quantity of carotenoids extracted from two passion fruit varieties. As reported for orange juice, the authors also observed increase in the concentration of total carotenoids extracted (~9-27%) from passion fruit as the pressure treatments increased (*4*). Table I summarizes the results from different studies that have explored the effects of HHP processing on the extractability of carotenoids.

In the prior work that reports increase in bioactive molecules as an effect of pressure treatment processing, most of the authors attributed their observations to morphological changes in cellular organelles (mitochondria, cytoplasm, and vacuole), which resulted in the migration of intracellular constituents to the food matrix (*5*–*11*).

Table I. HHP Processing Effect on the Extractability of Carotenoids from Different Food Matrixes

HHP conditions	*HHP effect*	*Reference*
Orange juice		
400 MPa /40 °C/1 min	>53.88% total carotenoids	(*7*)
	>33.76% α-carotene	
	>30.24% β-carotene	
	>45.87% α-cryptoxanthin	
	>43.21% β-cryptoxanthin	
	>44.52% zeaxanthin	
	>75.43% lutein	
Orange juice		
100 MPa/60 °C/5 min	>10.03% total carotenoids	(*3*)
	>11.34% α-cryptoxanthin	
	>32.53% zeaxanthin	
350 MPa/30 °C/2.5 min	>24.11% total carotenoids	
	>41.03% zeaxanthin	
	>29.57% β-carotene	
	>16.90% α-carotene	
	>31.93% β-cryptoxanthin	
	>18.44% lutein	
400 MPa/40 °C /1 min	>31.73% total carotenoids	
	>59.26% zeaxanthin	
	>43.50% β-carotene	
	>37.51% α-carotene	
	>28.92% β-cryptoxanthin	
	>23.47% lutein	
Carrots, green beans and broccoli		
400 and 600 MPa/2 min	= total carotenoids	(*36*)
	≥14.00% carotenoids bioavailability	
Mango pulp		
600 MPa/4 °C/3 min	= total carotenoids	(*9*)
Avocado pulp		
600 MPa/4°C/3 min	>56.00% total carotenoids	(*10*)

Continued on next page.

Table I. (Continued). HHP Processing Effect on the Extractability of Carotenoids from Different Food Matrixes

HHP conditions	*HHP effect*	*Reference*
	>513.00% nexanthin b	
	>312.00% α-cryptoxanthin	
	>284.00% α-carotene	
	>220.00% β-cryptoxanthin	
	>40.00% lutein	
Papaya slices cv. Sunrise		
400 MPa/25 °C/ 1min	>156.00% total carotenoids	(*37*)

At present, limited information exists regarding the effects of HHP on the integrity of plant cell organelles (*12*). However, the effects of pressurization on other eukaryotic cells including animal and human have been reported (*13*); those results will be described in the following paragraphs in order to explain the effects of pressure intensities on cell integrity.

Depending on the pressure intensity applied, HHP treatments can be classified as physiological pressure (pHHP) and non-physiological HHP (non-pHHP) processes (*13*). For discussion purposes in this work, pressure treatments in the range of 0.1 to 100 MPa are defined as pHHP, whereas processes with pressures above 100 MPa are classified as non-pHHP processes. It has been observed that pHHP treatments induce morphological alterations in cellular organelles and evoke a reversible stress similar to that observed for heat shock treatments (*13*). Therefore, pHHP treatments cause reversible alterations and normally do not affect cellular viability. Further, it has been noted that non-pHHP (>100 MPa) treatment effects on eukaryotic cells depend on the cell type. HHP treatments <150 MPa does not affect cellular viability in human cells, but generate apoptosis in murine cells (*13*). In yeast, non-pHHP treatment above 200 MPa cause damage in cell wall and subcellular structure, mainly in nucleus and mitochondria (*14*).

The effects of high pressure on vegetable structures has been previously studied by Préstamo and Arroyo (*15*). In their observations a HHP treatment of 400 MPa/30 min/5 °C, modified the permeability of plant cells and enabled migration of metabolites from inside to outside the cell (*15*). Additionally, Hartmann *et al.* (*16*) evaluated the effects of HHP treatments (55-120 MPa/3 h) on chloroplasts of two types of *Spirogyra* algae. Authors observed that pressure treatments of 120 MPa/2 min modified the structure of the chloroplast; changes that were only observed in 30% of the chloroplasts for samples treated at a lower pressure 55 MPa/30 min (*16*).

Thus, the aim of the work presented in this chapter was to evaluate the stability of carotenoid profiles from avocado, papaya and mango pulps treated at non-pHHP processing conditions (>100 MPa).

Materials and Methods

Fruit Samples and Chemicals

Avocados (*Persea americana* Mill, cv. Hass) used in the study were obtained from the Mexican region of Michoacan. Mangoes (*Mangifera indica* cv. Tommy Atkins) were obtained from Mexican regions of Chiapas and papayas (*Carica papaya*) were harvested at the Mexican region of Cohetzala, Puebla. All-*trans*-β-carotene and β-apo-8′carotenal were obtained from Sigma-Aldrich Co. (St. Louis, MO., USA). All-trans-lutein, all-trans-β-cryptoxanthin and zeaxanthin were purchased from Indofine Chemical Co. (Hillsborough, NJ., USA). Other carotenoid standards were purchased from Carotenature (Lupsingen, Switzerland).

Sample preparation varied slightly depending on the fruit used, but in general fruits were washed, sanitized, peeled and macerated into paste for avocado and papaya samples. Mango was cut in to cubes (~2.5 cm^2). Prior to packaging, the samples were vacuum de-aerated (-88.2 kPa) with an Ultravac UV2100A pump (Koch Equipment LCC, Kansas, City, MO, USA). Fruit samples (200 g) were vacuum packaged into oxygen impermeable plastic bags (14 cm length x 12.6 cm width x 3 cm height) of a thickness of 0.018 cm. After packaging samples were placed in cold water (1-3°C for 20 min), prior to processing, to reach an internal temperature of ~5°C.

Processing Equipment and Conditions

Avocado and mango samples were processed at 600 MPa for 3 min, using a 215L ULTRA HHP processing unit (Avure Technologies, Kent, WA, USA). The time required to reach 600 MPa was 3.5 min. The decompression time was 2.75 min. Purified water was used as pressurization media. Samples were pressurized at a vessel temperature of 23°C. Papaya samples were processed in a laboratory scale high hydrostatic pressure processing unit Model 2L-700 from Avure Technologies (Vasteras, Sweden). A full factorial experimental design was applied to papaya samples, that included three different pressure intensities (400, 500 and 600 MPa), two processing times (1 and 3 min), and two processing temperatures (25 and 40 °C).

Carotenoid Analyses

Carotenoid extraction and analyses were performed by high performance liquid chromatography with photodiode array detector (HPLC-PDA) according to the procedure described by Jacobo-Velazquez and Hernandez-Brenes (2012) (*10*). The HPLC system used was composed of two 515 binary pumps, a 717-plus autosampler, and a 996-photodiode array detector (Waters Corp., Mildford, MA). Carotenoids were separated on a 4.5 mm x 250 mm, 5 μm, C-30 reverse-phase column (YMC carotenoid, Wilmington, USA). The mobile phases consisted of methanol/MTBE/water (81:15:4, v:v, phase A) and MTBE/methanol/water

(90:6:4, v:v, phase B). The gradient solvent system was 0/100, 90/25, 100/100, 140/100 (min/% phase A) at a flow rate of 1 mL/min. Chromatographic data was processed with the Millennium software v3.1 (Waters Corp., Mildford, MA). Carotenes and xanthophylls were quantified at 450 nm and precursors at 286 and 350 nm.

Results and Discussion

Effect of High Hydrostatic Pressure Treatment on Carotenoid Contents of Various Tropical Fruits

Avocado cv. Hass Paste

Avocados (*Persea americana*) are an important source of bioactive molecules. The main antioxidants present in the fruit include xanthophylls (mainly lutein), vitamin C, vitamin E, and persenone A and B (*17–20*). To achieve shelf-life extension of avocado fruit, HHP is possibly the most effective technology, resulting in an excellent product-technology match (*21–24*). The ability of HHP to preserve the fresh-like characteristics of avocado led to the expansion of the technology to industrial scale, and at the present, avocados are being HHP processed and consumed by humans worldwide. One of the objectives of the present work was to evaluate the effects of HHP processing, under commercial conditions (600 MPa/3 min), on the stability of avocado carotenoids by HPLC-PDA and to compare it with the stability of other tropical fruits in subsequent sections of this chapter.

Carotenoids identified in the unprocessed avocado paste included lutein, β- and α- carotene, β- and α- cryptoxanthin, luteoxanthin isomers, neoxanthin, zeaxanthin, and two other possible xanthophylls, profiles that agree with those reported by Jacobo-Velazquez and Hernandez-Brenes (*10*). The individual carotenoid present at the highest concentration in avocado pulp was lutein (~56% of total carotenoids). Results that were consistent with previous reports on the characterization of the main carotenoids present in Hass avocados (*10*, *25*). Concentrations of total carotenoids in HHP processed samples were ~56% higher when compared to unprocessed samples (~38 ppm). The individual carotenoid that showed the highest percent increase in concentration was neoxanthin b (513%), followed by α- cryptoxanthin (312%), α-carotene (284%), β-cryptoxanthin (220%), β-carotene (107%) and lutein (40%).

Papaya cv. Maradol Paste

Attractive color of papaya fruit has been mainly attributed to its carotenoid content (*26*). In the work discussed herein the commercially relevant papaya cultivar Maradol was used, which was developed in Cuba, and is distinguished by a red-orange skin, a salmon-red pulp, and a large size (1.5-2.6 kg) (*27*). Papaya fruit has a short shelf-life and thus it is a good candidate for further processing to improve its stability (*28*). Prior work on papaya processing has been able to generate information on the difficulties involved in the preservation of the fresh fruit organoleptic characteristics (*29*). Papaya is highly susceptible to quality losses during thermal processing, including appearance of undesirable flavors (*26–30*). In most papaya processed products, the addition of other fruits to the formulation has been required in order to mask the undesirable flavors generated by thermal processing (*31*).

In the HHP papaya processing study reported herein three different pressure intensities (400, 500 and 600 MPa), were used, also two processing times (1 and 3 min), and two processing temperatures (25 and 40 °C). Results indicated that HHP treatments did not influenced chromatographic carotenoid profiles of the papaya pulp (Tables II and III). However, quantitative changes in the profiles induced by HHP treatments were observed (Table II). In order to describe general trends in Table II carotenoids were grouped as precursors, xanthophylls, carotenes, and total carotenoids. It was observed that the 600 MPa/3 min HHP treatments resulted in increase of total carotenoids contents of papaya puree that were 154% higher than the levels found in the unprocessed control samples. Similar increase have been reported for other macerated fruit matrixes such as avocado puree (156% of increase to 600 MPa/3 min) (*10*), persimmon puree (116% of increase to 50-400 MPa/15 min) (*4*) and orange juice (154% of increase to 400 MPa/1 min/40 °C) (*7*, *10*, *30*).

Results from the papaya processing study indicated that increase in pressure intensity resulted in the most significant modifications on carotene, xanthophyll, and carotenoid contents of papaya puree. Overall data analysis of main effects indicated that total carotenoid concentrations (Figure 1) and all three carotenoid agruppations (Figure 2) were significantly affected ($p < 0.05$) by the pressure intensity and temperature treatments applied, while the processing time treatment was not significant ($p > 0.05$). Therefore Figures 1 and 2 are main effect graphic representations (interaction plots) of the treatment effects that were significant, and exclude the processing time effect. Overall, HHP treatments in combination with the 40 °C processing temperature increased total carotenoid contents by 75% (Figure 1). Results were in agreement with prior work that evaluated the effects of mild thermal treatments on the carotenoid content of canned lettuce, carrot, spinach and beans; in which the authors reported increase in the concentrations of total carotenoids that ranged from 83-139% in reference to control treatments (*32*). Observations that have been attributed in the prior literature to thermally induced release of carotenoids from their cellular locations in protein complexes, therefore improving their extractability (*32*).

Interestingly, in the present study the concentration of total carotenoids for samples pressurized at 40 °C was not affected by the pressure intensity. However, for samples that were pressurized at room temperature (25 °C) a direct correlation between pressure intensity and total carotenoid concentrations was observed (Figure 1); a trend that was also observed when total carotenoids (processed at 25 °C) were broken down into the groups of precursors and carotenes (Figure 2A and B).

Table II. HHP Processing and Temperature Effects on the Extractability of Carotenoids from Papaya Puree

Carotenoid group		*Concentration (mg kg^{-1} papaya cv. Maradol puree)*[i]															
		25 °C								*40 °C*							
		NO HHP		*400 MPa*		*500 MPa*		*600 MPa*		*NO HHP*		*400 MPa*		*500 MPa*		*600 MPa*	
Precursors[ii]	1 min	13.41	f	30.54	abcd	28.36	bcde	37.63	ab	29.28	abcd	19.00	def	40.78	a	15.73	f
	3 min	13.41	f	25.60	bcdef	31.46	abc	29.70	abcd	24.60	cdef	22.83	cdef	16.85	ef	18.52	def
Xanthophylls	1 min	6.71	d	10.47	cd	10.93	cd	18.81	a	11.79	bcd	12.28	bcd	14.42	abc	10.60	cd
	3 min	6.71	d	9.84	cd	17.17	ab	9.49	cd	7.87	d	18.84	a	11.88	bcd	9.06	cd
Carotenes	1 min	28.15	g	32.20	efg	74.36	a	42.86	bcdefg	37.41	cdefg	41.29	bcdefg	59.90	ab	50.74	bcde
	3 min	28.15	g	30.90	fg	37.81	cdefg	49.40	bcdef	52.08	bcd	55.00	bc	34.81	defg	57.54	ab
Total carotenoids	1 min	48.27	e	73.21	bcde	113.64	a	99.30	ab	78.48	bcde	72.57	bcde	115.10	a	77.08	bcde
	3 min	48.30	e	66.34	cde	86.44	abcd	88.59	abcd	84.54	abcd	96.67	abc	63.53	de	85.12	abcd

[i] All values are means of three independent determinations. [ii] Values with different letters in the same carotenoid group are statistically significant ($p<0.05$).

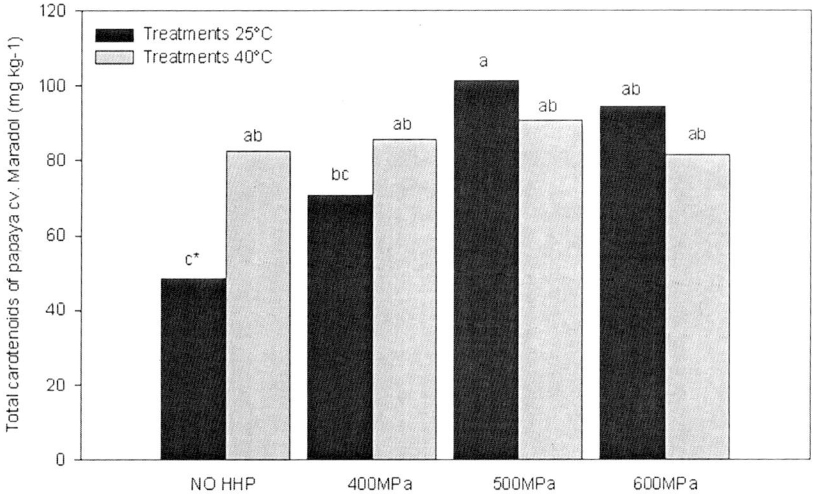

*Figure 1. Effects of HHP processing and temperature on the extractability of total carotenoid from papaya cv. Maradol puree. *Bars with different letters indicate significant differences ($p<0.05$).*

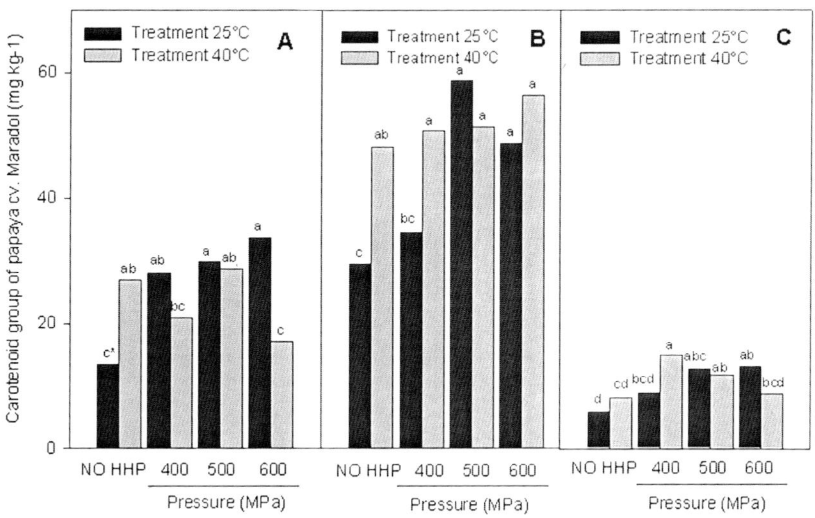

*Figure 2. Effects of the pressure intensity and temperature interaction on the extractability of papaya cv. Maradol carotenoids (A) Precursors; (B) Carotenes; (C) Xanthophylls. *Bars with similar letters for the same carotenoid group indicate that treatments are not significantly different ($p>0.05$).*

Mango Cubes cv. Tommy Atkins

As previously discussed for papaya and avocado, the HHP stability of mango carotenoids was also evaluated. Mangoes were selected because they are also a tropical fruit that contains high carotenoid concentrations. For the mango processing study the effect of HHP processing under commercial conditions (600 MPa/3 min) on the stability of mango carotenoids was determined. The chromatographic profile of mango carotenoids for the cv. Tommy Atkins can be observed in Figure 3. The tentative identification and quantification of the compounds is then presented in Tables III and IV.

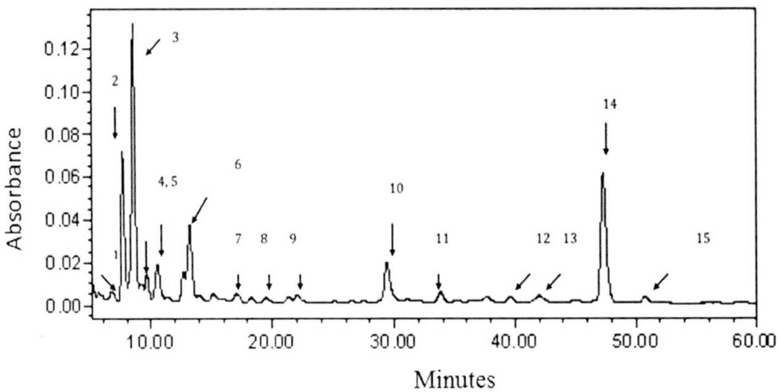

Figure 3. Separation of carotenoids from processed mango with high hydrostatic pressure by HPLC-DAD (450 nm). (1) Luteoxanthin isomers; (2) Violaxanthin; (3) Violaxanthin; (4)Luteoxanhin isomer; (5) Luteoxanthin isomer; (6) 9-cis-violaxanthin; (7) Luteoxanthin isomer; (8) Not identified; (9) 13-cis-violaxanthin; (10) Internal standar (apo-β-8′-carotenal); (11) Not identified; (12) all-trans-β-cryptoxanthin; (13) cis-β-cryptoxanthin; (14) all-trans-β-carotene; (15) 9-cis-β-carotene.

Qualitative and quantitive carotenoids profiles obtained in the present study are in agreement with prior publications on the topic (*33*, *34*). The main mango carotenoids identified were *all-trans*-β-carotene and violaxanthin. Presence of the 13-cis-β-carotene isomer, previously reported for thermal processed mango products (*35*), was not observed in the present work. Results indicated that HHP application did not affect the concentration of total carotenoids (~96 ppm) initially present in mango pulp. Therefore, HHP proved to be an effective treatment for the pasteurization of the mango pulp without the destruction of these bioactive molecules. One of the main advantages of the application of HHP for the stabilization of mango, when compared with other thermal processing technologies, was that violaxanthin was not degraded (Table IV). Other processing technologies such as drying degrade violaxanthin, producing unstable diepoxides derivatives (*34*). Likewise, the formation of violaxanthin diepoxides has been reported for mango juice processed by thermal processing (*32*).

Table III. Identifications of Carotenoids Present in Chromatographic Profiles of Mango (*Mangifera indica* cv. "Tommy Atkins") Pulp Processed with High Hydrostatic Pressure at 600 MPa/3 min to 4°C

Peak no. (retention time, min)[a]	λ_{max} *(nm)*[b]	*Identification*	*Identification method*[c]
1 (6.7)	399.6, 421.3, 445.5	luteoxanthin isomers	A,C
2 (7.6)	416.5, 440.7, 469.5	violaxanthin	A,C
3 (8.6)	416.5, 440.7, 469.5	violaxanthin	A,C
4 (9.5)	397.2, 421.3, 447.9	luteoxanthin isomers	A
5 (10.6)	399.6, 421.3, 447.9	luteoxanthin isomers	A
6 (13.2)	412.0, 335.8, 464.9	9-cis-violaxanthin	A,C
7 (15.2)	400.0, 421.3, 447.9	luteoxanthin isomers	A,C
8 (18.5)	412.0, 438.0, 467.0	not identified	A
9 (22)	410, 435.8, 464.9	13-cis-violaxanthin	A,C
10 (29)	467.3	apo-β-8′-carotenal	B
11 (33.6)	416.5, 440.7, 472.1	not identified	A
12 (37.3)	(423.7), 445.0, 470.0	all-trans-β-cryptoxanthin	A,C
13 (40.4)	(420.0), 452.7, 479.4	cis- β-cryptoxanthin	A,C
14 (45.6)	(422.0), 452.7, 472.1	all-trans-β-carotene	A, B,C
15 (48.6)	424.0, 445.0.5 472.1	9-cis-β-carotene	A,C

[a] Peak number in according with the chromatogram shown in Figure 3. [b] λ_{max} obtained by HPLC-DAD, value in parentheses indicates shoulder instead of peak. [c] Methodology used to identify carotenoids: (A) Extraction of the UV-visible spectrum (λ_{max}) and comparison with data available in literature (*35*, *38*, *39*); (B) Identification by direct comparison of retention time and UV-vis absorption spectra with commercial standards; (C) Comparison of elution order and approximation of retention time with information available in literature under similar chromatographic conditions on the C_{30} column (*39*).

Table IV. Quantification of Carotenoids Present in Chromatographic Profiles of Mango (*Mangifera indica* cv. "Tommy Atkins") Pulp Processed with High Hydrostatic Pressure at 600 MPa/3 min to 4°C

Peak no.(retention time, min)[a]	*Identification*	*Concentration (mg kg⁻¹)*[b] *Unprocessed*		*Processed*	
1 (6.7)	luteoxanthin isomers	3.45	a[c]	3.61	a
2 (7.6)	violaxanthin	6.04	a	3.77	a
3 (8.6)	violaxanthin	9.19	a	4.57	a
4 (9.5)	luteoxanthin isomers	3.45	a	3.61	a
5 (10.6)	luteoxanthin isomers	5.60	a	3.55	a
6 (13.2)	9-cis-violaxanthin	6.38	a	3.61	a
7 (15.2)	luteoxanthin isomers	3.79	a	3.63	a
8 (18.5)	not identified	3.76	a	3.70	a
9 (22)	13-cis-violaxanthin	3.77	a	3.55	a
10 (29)	not identified	3.56	a	3.57	a
11 (33.6)	all-trans-β-cryptoxanthin	3.54	a	3.73	a
12 (37.3)	cis-β-cryptoxanthin	3.73	b	4.52	a
13 (40.4)	all-trans- β-carotene	8.86	a	9.03	a
14 (45.6)	9-cis-β-carotene	3.66	a	4.16	a

[a] Peak number according to elution order of the chromatogram (Figure 3.; [b] Concentrations were expressed as β-carotene equivalents and determinate to 450nm. [c] Values with different letters for the same compound indicate significant difference ($p<0.05$).

Conclusions

HHP processing of tropical fruits can be used as a strategy to retain and/or increase the carotenoid content of tropical fruits such as avocado, papaya, and mango. For papaya and avocado pulps, higher concentrations of total carotenoids were extracted after HHP processing. Likewise, the differences between the concentration of total and individual carotenoids of mango pulp before and after processing were not significant ($p<0.05$). Further studies are needed to better understand the physiological effects of the application of high pressure on plant cells, since its application causes oxidative stress within plant tissues, and RNA molecules may remain intact in pressurized samples suggesting that the cells are metabolically active and perhaps capable of biosynthesizing stress related metabolites. Although at the non-physiological pressures evaluated in the present study, carotenoid increases induced by HHP processing are possibly related to their increased extractability, which could have a positive impact on their absorption during human digestion. However pharmacokinetic studies with humans or digestion models using HHP processed fruits are needed in order to further validate the potential health benefits of carotenoids from tropical fruits preserved by the technology.

References

1. Polydera, A. C.; Stoforos, N. G.; Taoukis, P. S. *Int. J. Food Sci. Technol.* **2004**, *39*, 783–791.
2. Pandrangi, S.; Balasubramaniam, V. M. In *Emerging Technologies for Food Processing*; Da-Wen, S., Ed.; Elsevier: London, 2005; pp 33–45, ISBN 10: 0-12-676757-2.
3. Sánchez-Moreno, C.; Plaza, L.; De Ancos, B.; Cano, M. P. *J. Agric. Food Chem.* **2003**, *51*, 647–653.
4. De Ancos, B.; González, E.; Cano, M. P. *J. Agric. Food Chem.* **2000**, *48*, 3542–3548.
5. Shimada, S.; Andou, M.; Naito, N.; Yamada, N.; Osumi, M.; Hayashi, R. *Appl. Microbiol. Biot.* **1993**, *40*, 123–131.
6. Cheftel, J. C. *Food Sci. Technol. Int.* **1995**, *1*, 75–90.
7. Sánchez-Moreno, C.; Plaza, L.; Elez-Martínez, P.; De Ancos, B.; Martín-Belloso, O.; Cano, M. P. *J. Agric. Food Chem.* **2005**, *53*, 4403–4409.
8. Torres, J. A.; Velázquez, G. *J. Food Eng.* **2005**, *67*, 95–112.
9. Ramos-Parra, P. M.S. Thesis, Tecnológico de Monterrey–Campus Monterrey, Monterrey, N.L. Mexico, 2006.
10. Jacobo-Velázquez, D. A.; Hernández-Brenes, C. *Innovative Food Sci. Emerging Technol.* **2012**, *16*, 121–128.
11. Villarreal-Lara, R. M.S. Thesis, Tecnológico de Monterrey–Campus Monterrey, Monterrey, N.L. Mexico, 2011.
12. Gonzalez, M. E.; Barrett, D. M. *J. Food Sci.* **2010**, *75*, 12–130.
13. Frey, B.; Janko, C.; Ebel, N.; Meister, S.; Schlucker, E.; Meyer-Pittroff, R.; Fietkau, R.; Herrmann, M.; Gaipl, U. S. *Curr. Med. Chem.* **2008**, *15*, 2329–2336.
14. Sato, M.; Kobori, H.; Shimada, S.; Osumi, M. *FEMS Microbiol. Lett.* **1995**, *131*, 11–15.
15. Préstamo, G.; Arroyo, G. *J. Food Sci.* **1998**, *63*, 1–4.
16. Hartmann, M.; Kreuss, M.; Sommer, K. *Cell. Mol. Biol.* **2004**, *50*, 479–484.
17. Heinonen, M. I.; Ollilainen, V.; Linkola, E. K.; Varo, P. T.; Koivistoinen, P. E. *J. Agric. Food Chem.* **1989**, *37*, 655–659.
18. Lu, Q. -Y.; Arteaga, J. R.; Zhang, Q.; Huerta, S.; Go, V. L. W.; Heber, D. *J. Nutr. Biochem.* **2005**, *16*, 23–30.
19. Bergh, B. *California Avocado Soc. Yearb.* **1992**, *76*, 123–135.
20. Kim, O. K.; Murakami, A.; Nakamura, Y.; Takeda, N.; Yoshizumi, H.; Ohigashi, H. *J. Agric. Food Chem.* **2000**, *48*, 1557–1563.
21. López-Malo, A.; Palou, E.; Barbosa-Cánovas, G. V.; Welti-Chanes, J.; Swanson, B. G. *Food Res. Int.* **1998**, *31*, 549–556.
22. Jacobo-Velázquez, D. A.; Hernández-Brenes, C. *J. Food Sci.* **2010**, *75*, S264–S270.

23. Jacobo-Velázquez, D. A.; Ramos-Parra, P. A.; Hernández-Brenes, C. *J. Food Sci.* **2010**, *75*, S286–S291.
24. Jacobo-Velázquez, D. A.; Hernández-Brenes, C. *J. Food Sci.* **2011**, *76*, S388–S395.
25. USDA-NCC, 1998. http://www.nal.usda.gov/fnic/foodcomp/Data/Carot/index.html.
26. Cano, M. P.; De Ancos, B.; Lobo, M. G.; Monreal, M. *J. Sci. Food Agric.* **1996**, *71*, 351–358.
27. Basulto, F. S.; Duch, E. S.; Gil, F. E . Y.; Plaza, R. D.; Saavedra, A. L.; Santamaría, J. M. *Interciencia* **2009**, *34*, 583–588.
28. Gouado, I.; Schweigert, F. J.; Ejoh, R. A.; Tchouanguep, M. F.; Camp, J. V. *Eur. J. Clin. Nutr.* **2007**, *61*, 1180–1188.
29. Parker, T. L.; Esgro, S. T.; Miller, S. A.; Myers, L. E.; Meister, R. A.; Toshkov, S. A.; Engeseth, N. J. *Food Chem.* **2010**, *118*, 861–869.
30. Argaiz, A.; López-Malo, A. *Rev. Esp. Cienci. Tecnol. Aliment.* **1995**, *35*, 92–100.
31. Tiwari, RB Studies on blending of guava and papaya pulp for RTS beverage. *Indian Food Packer* **2000**, *54*, 68–72.
32. Dietz, J. M.; Kantha, S.; Erdman, J. W., Jr. *Plant Foods Hum. Nutr.* **1988**, *38*, 333–341.
33. Mercadante, A.; Rodríguez-Amaya, D.; Britton, G. *J. Agric. Food Chem.* **1997**, *45*, 120–123.
34. Mercadante, A.; Rodríguez-Amaya, D. *J. Agric. Food Chem.* **1998**, *46*, 128–130.
35. Pott, I.; Marx, M.; Neidhart, S.; Muhlbauer, W.; Carle, R. *J. Agric. Food Chem.* **2003**, *51*, 4527–4531.
36. McInerney, J. K.; Seccafien, C. A.; Stewart, C. M.; Bird, A. R. *Innov. Food Sci. Emerg.* **2007**, *8*, 543–548.
37. Sánchez-Moreno, C.; De Ancos, B.; Plaza, L.; Elez-Martínez, P.; Cano, M. P. *Cr. Rev. Food Sci. Nutr.* **2009**, *49*, 552–576.
38. Chen, B. H.; Peng, H. Y.; Chen, H. E. *J. Agric. Food Chem.* **1995**, *43*, 1912–1918.
39. Meléndez-Martínez, A. J.; Britton, G.; Vicario, I. M.; Heredia, F. J. *J. Agric. Food Chem.* **2005**, *53*, 6362–6367.

Chapter 3

Climate and Salinity Effects on Color and Health Promoting Properties in the Pomegranate (*Punica granatum* L.) Fruit Arils

H. Borochov-Neori,*,[1] N. Lazarovitch,[2] S. Judeinstein,[1] B. S. Patil,[3] and D. Holland[4]

[1]Southern Arava Research and Development, Hevel Eilot, 88820, Israel
[2]Ben-Gurion University of the Negev, Wyler Department of Dryland Agriculture, French Associates Institute for Agriculture and Biotechnology of Drylands, Jacob Blaustein Institutes for Desert Research, Sede-Boqer Campus, 84990, Israel
[3]Texas A&M University, Department of Horticultural Science, Vegetable and Fruit Improvement Center, College Station, Texas, 77843-2119, United States
[4]Agricultural Research Organization, Unit of Deciduous Fruit Trees Sciences, Newe Ya'ar Research Center, Ramat Yishay, 30095, Israel
***E-mail: hamutalborochov@rd.ardom.co.il. Tel: +972-8-6355747. Fax: +972-8-6355730.**

Pomegranate (*Punica granatum* L.) has recently become a commercially important crop due to multitude health promoting properties attributed to different parts of the fruit. Worldwide pomegranate production has expanded greatly in traditional and new locations of diverse climatic and water conditions. The pomegranate industry demands intensely red fruit with high antioxidant content. It is therefore especially important to investigate the effects of climate and water quality on pomegranate anthocyanin and phenolics accumulation. Our research has focused on the edible part of the fruit, specifically, the arils. A diverse selection of pomegranate cultivars was employed. To explore climate effect, fruit that developed and ripened under a wide range of temperature regimes were studied. To explore water quality effect, fruit from plants irrigated with a wide range of salinities, 0.5-9 dS m^{-1}, were

© 2013 American Chemical Society

studied. Anthocyanins were analyzed by RP-HPLC and phenolics content was determined by the Folin-Ciocalteau assay. Both abiotic factors considerably affected arils' composition. Cooler temperatures enhanced both pigment and antioxidant accumulation. Climate also affected anthocyanin composition. Increased salinity enhanced phenolics accumulation in both accessions but reduced that of anthocyanins in a cultivar dependent manner. Our results can benefit breeding and agricultural efforts to enhance pomegranate fruit quality, especially, in face of global warming and water quality deterioration.

Keywords: anthocyanins; anthocyanin accumulation; arils; climatic conditions; phenolics; Pomegranate; *Punica granatum* L.; salinity

Introduction

Pomegranate (*Punica granatum* L.) has recently become an economically important crop due to multitude health promoting properties attributed to the fruit and its products by numerous *in vivo* and *in vitro* studies [extensively reviewed in (*1*, *2*), (*3*, *4*)]. Pomegranate commercialization has greatly increased (*5*), leading to the expansion of worldwide production in both traditional and new locations of diverse climatic and water conditions (*6*).

The pomegranate health qualities were attributed to the exceptionally high antioxidative capacity of the fruit (*7*). The latter is probably effected by the high content and unique composition of soluble phenolic compounds, especially in the fruit peel (*7–11*). Pomegranate fruit are usually consumed fresh, the edible part being the fruit arils, or as processed products, mainly juice. Commercial pomegranate juice (PJ) contains significant amounts of polyphenols from the fruit peel and exhibits a high antioxidative capacity (*7*, *9*). PJ also contains anthocyanins that originate mostly from the arils and impart its characteristic color (*12–14*). The consumer seeks in the fruit and its products both the distinctive intense red color and the high health value; thus, the pomegranate industry requires red colored fruit with high phenolics' content.

The soluble phenolics in the pomegranate arils include anthocyanins, hydrolysable tannins (mostly ellagitannins and small amounts of gallagyl esters and gallotannins), hydroxybenzoic acids (including gallic acid, protocatechuic and vanilic acid derivatives) and hydroxycinnamic acids (mostly caffeic, ferulic and coumaric acid derivatives) (*7*, *15*). Six pigments are common to the pomegranate fruit arils, including 3-monoglucosides and 3,5-diglucosides of cyanidin, delphinidin and pelargonidin (*12*, *15–17*). All six anthocyanins are present in fruit from Spanish, Californian, Tunisian, Italian, Iranian, Uzbekistanian, Peruvian and Israeli accessions (*12–15, 18–20*). The quantities and proportions of the individual pigments depend on variety and environmental and cultural variables. Recently, additional minor anthocyanins were identified in the arils of Peruvian pomegranates, including cyanidin-pentoside-hexoside and cyanidin-pentoside (*15*).

Phenolics' accumulation in plants, specifically that of flavonoids, including anthocyanins, is sensitive to environmental conditions and may be induced by a number of abiotic stress factors, such as strong light, UV-B radiation, extreme temperatures, drought and osmotic stress (*21–27*). The responses, however, are crop, variety and organ dependent.

Cool temperatures enhance and high temperatures reduce anthocyanin accumulation (*28–30*) probably due to their adverse effects on biosynthesis and degradation rates. We have previously reported on the effects of seasonal (*20, 31*) and geographical (*32*) variations on color, anthocyanin and phenolics accumulation in pomegranate fruit of a diverse collection of accessions grown in Israel. The external and internal color and anthocyanin accumulation in fruit that matured and ripened under extreme hot temperatures were lower compared to moderate climate conditions (*20, 31, 32*). Seasonal variations also affected arils' anthocyanin composition (*20*). Climatic effects on the concentration of total soluble phenolics in the pomegranate arils were not fully evaluated (*20, 31, 32*).

Differences in irrigation water salinity and soil properties may have contributed to our previously reported geographical variations in anthocyanin and phenolics accumulation in the pomegranate fruit (*32*). Pomegranate is considered to be moderately tolerant to salinity (*33, 34*). Studies on ion uptake and translocation as well as sugar accumulation in the leaves of three Iranian pomegranate cultivars under increasing salinity levels suggested a reasonable tolerance up to 40 mM NaCl for all three varieties (*35*); the effect of salinity on growth rate, however, was cultivar dependent (*36*). Considerable variations in salinity tolerance among varieties were reported in a comparative study on ten Iranian pomegranate cultivars (*37*). A recent study on the evapotranspiration, crop coefficient and growth in young plants of two pomegranate varieties suggested that the crop may possibly be listed as moderately sensitive rather than moderately tolerant to salinity (*38*). Arils' anthocyanin content in pomegranate fruit of 'Mollar' cultivar was higher when grown in low salt, organic material rich compared to poor salty soil (*12*). To date, no detailed study exploring salinity effect on the accumulation of anthocyanins and phenolics in the pomegranate fruit has been reported.

The expansion of pomegranate production to new regions and the global trends of warming and water quality deterioration necessitate further understanding of climate and salinity effects on the fruit coloration and phenolics content to comply with market demands for intensely red antioxidant rich pomegranates. Knowledge of the factors and processes involved in anthocyanin and soluble phenolics accumulation can benefit the current breeding and agricultural efforts to enhance pomegranate fruit quality.

The present report describes studies on the effects of season temperatures and irrigation water salinity during pomegranate fruit development and ripening on anthocyanin and phenolics composition in the arils, i.e. the edible part, of the ripe fruit. A diverse selection of pomegranate cultivars, including deciduous and evergreen accessions, was employed. To explore climate effect, ripe fruit were harvested throughout the entire year, thus, being exposed to distinct temperature regimes during development and ripening, ranging from cool to high temperatures. To explore water salinity effect, fruit from plants irrigated with a wide range of salinities, from low to highly saline irrigation water, were studied. Arils were analyzed to establish temperature and salinity effects on 1) anthocyanin accumulation and composition, and 2) phenolics accumulation.

Materials and Methods

Plant Material

Climate Study

Ripe pomegranate (*Punica granatum* L.) fruit were collected from the pomegranate orchard of the southern Arava R&D experimental farm located in the Israeli southern Arava Valley (lat. 29°53′N; long. 35°3′E, desert climate). Trees were planted at a 3 x 5 m distance in Arava sandy loam soil (*Typic Torrifluvent*) (83% sand, 8% slit and 9% clay, containing 1.3% organic material). Year-round irrigation with local water (3 dS m^{-1}) was applied on a daily basis, according to local practice (12,000 m^3 water and 1 ton nitrogen $hectare^{-1}$). Horticultural management (pest control, pruning, thinning and harvesting) was based on recommended practices specified for southern Israel by the Ministry of Agriculture Extension Service. Three pomegranate accessions were studied: 'PG 128-29' (deciduous) and 'EG 1' and 'EG 2' (evergreen) (*20*).

Salinity Study

Plants of two deciduous pomegranate accessions, 'Wonderful' and 'SP-2', were grown in lysimeters situated at the Jacob Blaustein Institutes for Desert Research, Sede Boqer Campus, the Ben-Gurion University of the Negev, Israel, located in the heart of the Negev Desert (lat. 30°52′N; long. 34°46′E). A detailed description of the pomegranate lysimeters' orchard is given in Bhantana and Lazarovitch, 2010 (*38*). In short, single rooted cuttings were transplanted into 0.2 m^3 lysimeters, filled with Sede Boqer loess soil. Each lysimeter had a highly conductive drain filled with rock wool to control the matric potential at the bottom of the container. Salinity treatments began in 1-year old plants. The salinity of the irrigation water was brought to electrical conductivity (EC) of 0.5, 1.2, 3 , 6 and 9 dS m^{-1} with NaCl and $CaCl_2$ (in proportional concentrations to have an equivalent contribution to the solution EC) and fertilizer (Poly-Feed water soluble NPK fertilizer 14:7:28 + 1% MgO with micronutrient combination, Haifa Chemicals, Israel). Henceforth the treatments are denoted as EC 0.5, EC 1.2, EC 3, EC 6 and EC 9. Five plants of each accession, one per lysimeter and randomly distributed within the lysimeters' orchard, served as five replicates for each salinity level.

Fruit were sampled during two or three years on selected harvest dates. Both harvest dates and sampling years are specified in the corresponding sections under "RESULTS". On each sampling date, ripe fruit were selected by external criteria according to customary grower practices, including external color, size and shape. Fruit of a similar size were chosen from different trees and locations in the orchard and processed within 24h from harvest.

Meteorological Measurements

Temperature in the southern Arava R&D experimental farm was measured using electronic temperature sensor (MP101A-T7-W4W probe, Rotronic A.G., Switzerland).

Fruit Processing

Intact arils were separated by hand from the pith and carpellary membranes. Ripeness was further established by tasting; only non-astringent edible fruit were analyzed. Juice was prepared from the isolated arils by a solid fruit juice extractor (Juice Extractor, Model Le Duo, Magimix, France) and analyzed. Analytical assays were carried out in triplicates.

Juice Electrical Conductivity

Electrical conductivity (EC) in fruit arils' juice was measured by a conductometer (Metrohm, 644, Switzerland) equipped with 1-cm measuring cell and expressed in mS cm^{-1}.

Total Soluble Phenolic Content

Pomegranate aril juice was extracted with 80% methanol supplemented with 2mM NaF (1:2, v/v) and centrifuged (10,000 rpm for 10 min at 4°C, Sorvall Instruments RC5C, rotor no. SS-34). The supernatant was diluted 10-fold with double distilled water (DDW). Total soluble phenolics concentration was measured colorimetrically with Folin-Ciocalteau 2N phenol reagent (SIGMA Chemical Co, USA) according to Singleton and Rossi (*39*). Aliquots of 100 µL were added to 900 µL reaction solution consisting of 200 µL freshly prepared 10-fold diluted Folin-Ciocalteau reagent, 100 µL 20% Na_2CO_3 and 600 µL DDW. Pyrogallol (SIGMA Chemical Co, USA) was used for the calibration curve (0-100 µg mL^{-1}). The absorbance at 765 nm after 1-hour incubation was measured with a spectrophotometer (SHIMADZU Corporation, UV–1650PC, Kyoto, Japan), and the content of phenolics in the juice was expressed in pyrogallol equivalents, mg mL^{-1}.

RP-HPLC Analysis of Anthocyanins

Methanolic extracts of pomegranate arils' juice were prepared as described earlier. The supernatants were filtered through a 0.45µm PTFE filter before injection. Anthocyanin analysis was carried out as described in Borochov-Neori *et al.*, 2011 (*20*) with LaChrom Merck Hitachi HPLC system, consisting of pump Model L7100, Column oven L7350 and Mixer-Degasser L-7614, coupled with a diode array detector with 3D feature (Multiwavelength Detector, Jasco MD-2010 Plus), interface (Jasco LC-Net II / ADC) and scientific software (EZChrom *Elite*™ Client/Server version 3.1.6 build 3.1.6.2433). Twenty μL of extract were injected using a manual injector (Rheodyne, Rohnert Park, CA) and loaded onto a column of Lichrospher®100 RP-18 (5 μm particle size in 250x4 mm LichroCART® cartridge) with a guard column of the same packing material (4x4 mm LichroCART® cartridge). Column temperature was maintained at 40°C. The binary mobile phase consisted of phosphoric acid in DDW (0.1%, pH 2.4) (A), and acetonitrile (B). Elution was conducted as follows: 10% B at 1 ml/min for the first 10 min; 10% B to reach 20% by 15 min, at 1 ml/min; 20% B at 1 ml/min to reach 0.6 ml/min at 16 min; 20% B at 0.6 ml/min for 10 min. Following anthocyanin elution, the column was washed and equilibrated by 10-min post runs with 80% and 10% B, respectively.

Acetonitrile was HPLC grade (LiChrosolv®, Merck Chemicals); DDW was passed through a 0.20μm Nylon membrane. Phosphoric acid and NaF were of analytical grade.

Anthocyanin Identification and Quantification

Peak assignment was performed by the software on the basis of UV/VIS absorbance spectra and the retention times of anthocyanin standards. The standard library was constructed from delphinidin 3,5-diglucoside (D-3,5-G), cyanidin 3,5-diglucoside (C-3,5-G), pelargonidin 3,5-diglucoside (P-3,5-G), malvidin 3-glucoside chloride, delphinidin 3-glucoside (D-3-G), cyanidin 3-glucoside (C-3-G) and pelargonidin 3-glucoside (P-3-G) (Polyphenols Laboratories AS). Each standard (50-100 μg mL^{-1} methanol) was injected separately and the data acquired by the photodiode array detector with the 3D feature were incorporated into the system anthocyanin standard library. Under the conditions employed in this study, the retention time (in min) and wavelengths of UV/Vis maximal absorbance (in nm) for D-3,5-G, C-3,5-G, P-3,5-G, D-3-G, C-3-G and P-3-G were 4.1 and 274/520, 6.2 and 277/512, 9.4 and 274/497, 11.0 and 276/522, 16.3 and 279/515, 19.7 and 275/501, respectively. Relative standard deviation (RSD) for the retention times in repetitive runs was in the range of 0.4-1.9%.

Individual anthocyanins were quantified from the corresponding chromatogram peak area calculated by the software. Calibration curves were constructed with standards for each of the above six anthocyanins at four concentrations (0.01, 0.10, 0.25 and 0.50 mg mL^{-1}). RSD for peak areas in multiple runs was within 0.4-3.2%. Total anthocyanin concentration was calculated as the sum of concentrations of the individual anthocyanins.

Statistical Analysis

All data are reported as means ± their respective standard deviations (SD).

Results

Climate Effects on the Pomegranate Arils' Anthocyanin and Soluble Phenolics Composition

To study seasonal effects on the composition of anthocyanins and soluble phenolics in the pomegranate fruit arils we have analyzed fruit that ripened on different dates throughout the year employing both deciduous ('P.G. 128-29') and evergreen ('EG 1' and 'EG-2') accessions (*20*).

The three accessions are characterized by internally red colored fruit when produced under traditional cultivation conducts. 'P.G. 128-29' is an early ripening accession. Its harvest season in the southern Arava Valley starts at the beginning of July and can be extended until the end of December with the appropriate cultivation management (*31*). Fruit were sampled during mid-July and mid-December in the years 2005 and 2007. 'EG 1' and 'EG 2' accessions flower and produce fruit year round in the southern Arava Valley (M. Harari, unpublished results). Ripe fruit were sampled from December through August during the years 2006 and 2007 ('EG 1') and from July through January during the years 2005, 2007 and 2008 ('EG 2'). The analytical data from fruit of the three accessions were combined to construct whole year arrays of arils' anthocyanin and soluble phenolics composition and harvest dates/daily average temperatures (Figures 1-3). The range of daily average temperatures were 7-25°C (minimum) and 16-40°C (maximum). Data are presented as average and standard deviation for 5 – 7 fruit from the different sampling years.

Anthocyanin Content

Total anthocyanin content in the pomegranate arils is presented as a function of harvest date (Figure 1A) and the corresponding average daily minimal and maximal temperatures (Figure 1B). Anthocyanin accumulation largely varied with the sampling date (Figure 1A), being high in the winter, decreasing during spring and reaching very low levels in summer and autumn. Strong inverse correlation with harvest date temperatures was obtained (Figure 1B).

Anthocyanin Composition

The fruit arils of the three accessions contain the typical anthocyanins of pomegranate fruit, i.e. monoglucosides and diglucosides of delphinidin, cyanidin and pelargonidins (20), in proportions that are cultivar dependent. The content of each of the anthocyanins decreased with the progression in harvest season from winter to autumn; however, at different rates (data are not shown). Consequently, the relative abundance of the individual anthocyanins changed with season. Especially noticeable was the change in the proportions of mono- and di-glucosylated anthocyanins with harvest date (Figure 2A). The change markedly related to the average daily temperatures at harvest (Figure 2B). The majority of the anthocyanins were mono-glucosylated at cool temperatures and di-glucosylated at warm temperatures.

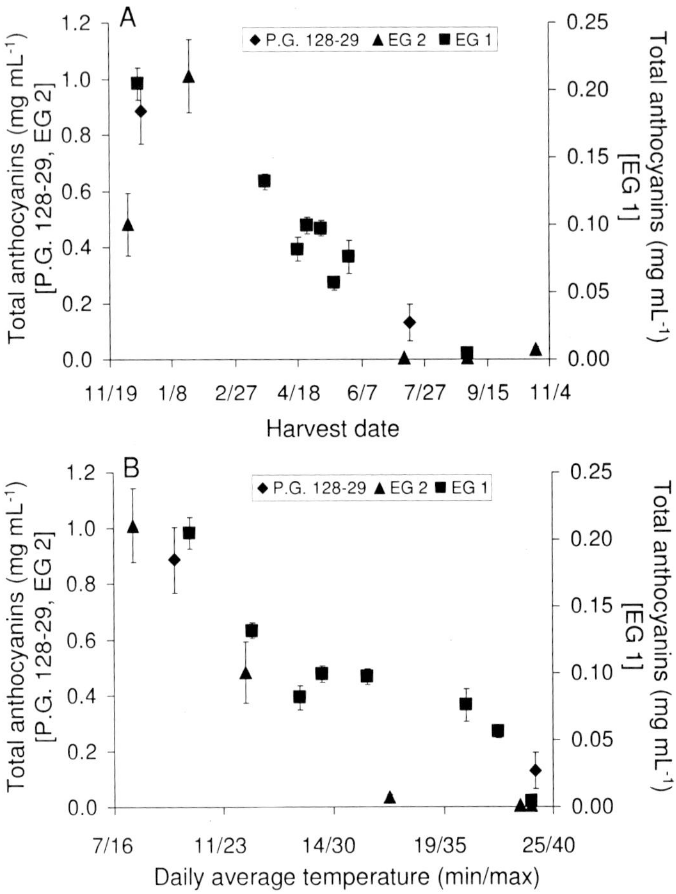

Figure 1. Arils' total anthocyanin content as a function of harvest date (A) and the corresponding average daily minimal and maximal temperatures (B).

Soluble Phenolics Content

The concentration of soluble phenolic compounds in the pomegranate fruit arils is presented as a function of harvest date (Figure 3A) and the corresponding average daily minimal and maximal temperatures (Figure 3B). The content of soluble phenolics was fairly stable during most of the year but increased significantly in fruit that ripened in the coolest season (~1-1.4 and ~1.6-1.8 mg mL^{-1} pyrogallol equivalents, respectively).

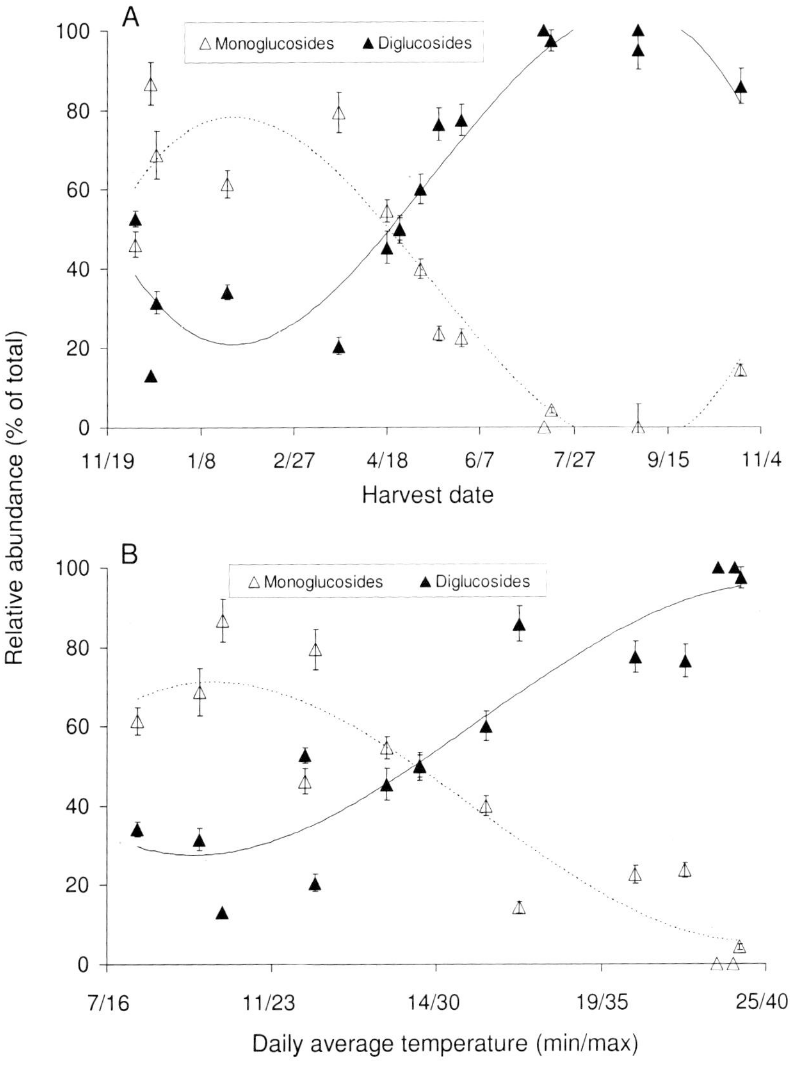

Figure 2. Relative abundance of mono- and di-glucosylated anthocyanins in the pomegranate arils as a function of harvest date (A) and the corresponding daily average minimal and maximal temperatures (B).

To summarize, anthocyanin accumulation and composition in the pomegranate fruit arils was cultivar dependent and highly sensitive to preharvest temperatures. Warmer temperatures were associated with marked reduction in pigment content and increase in the proportion of diglucosylated anthocyanins. The content of soluble phenolics was comparable in different accessions and remained stable during most of the year, but significantly increased in the coolest season.

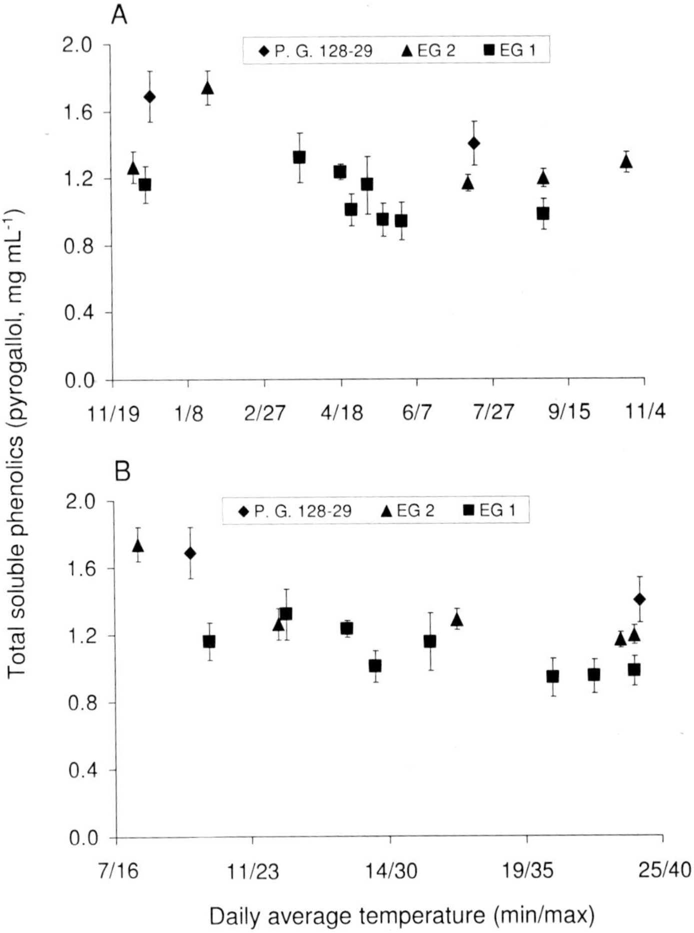

Figure 3. Total soluble phenolics content in the pomegranate arils as a function of harvest date (A) and the corresponding daily average minimal and maximal temperatures (B).

Salinity Effects on the Pomegranate Arils' Anthocyanin and Soluble Phenolics Composition

The effect of irrigation water salinity on the composition of anthocyanins and soluble phenolics in the pomegranate fruit arils was studied in two deciduous accessions: 'SP-2', an early ripening Turkmen accession, and 'Wonderful', a late ripening variety. Both accessions are characterized by red colored arils when produced under traditional cultivation conducts. 'SP-2' and 'Wonderful' fruit were collected for analysis at the end of August and the end of September, respectively, in the years 2009 and 2010.

Anthocyanin Content

Arils' anthocyanin content as a function of the irrigation water salinity is presented in Figure 4. 'Wonderful' arils were richer in anthocyanins compared to 'SP-2' arils at all salinity levels but more so at salinities up to EC 3. Pigment accumulation in 'Wonderful' fruit arils decreased markedly at the lower salinity levels and more moderately at salinities > EC 3. In 'SP-2' fruit arils, on the other hand, anthocyanin accumulation remained unaffected by salinities up to EC 6 and slightly decreased with irrigation at EC 9. It is note worthy that the EC measured in the arils' juice was nearly constant at all irrigation water salinities, 5.9±0.3 and 4.4±0.5 mS cm^{-1} in 'Wonderful' and 'SP-2', respectively.

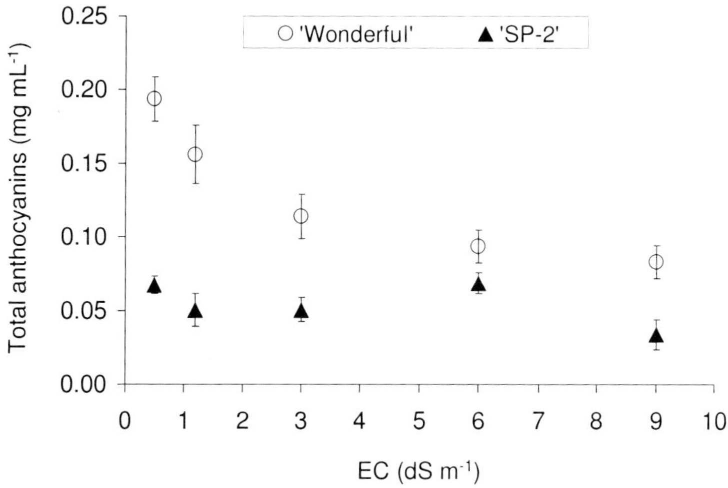

Figure 4. Effect of irrigation water salinity on arils' anthocyanin content in fruit of 'Wonderful' and 'SP-2' pomegranate accessions.

Anthocyanin Composition

The typical pomegranate anthocyanins were detected in fruit arils of the two accessions under all irrigation water salinities (data are not shown). The proportions of the individual anthocyanins in 'Wonderful' arils changed significantly with salinity. The relative abundance of C-3,5-G decreased from 53.6±2.7 to 40.7±2.0 % when irrigation water salinity was raised up to EC 3 and increased when the salinity was further elevated reaching 60.9±3.0% at EC 9.0; concomitant changes in the opposite direction were measured in the proportions of C-3-G and D-3-G. As a result, the relative level of diglucosides decreased considerably with salinity up to EC 3 and increased with further salinization (Figure 5). No clear salinity effect on the relative abundance of the individual anthocyanins in 'SP-2" arils could be established.

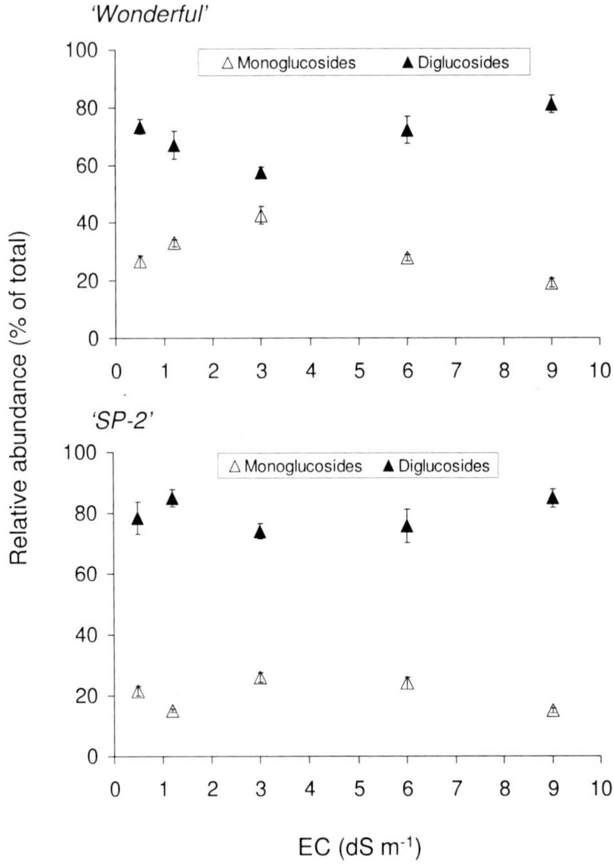

Figure 5. Effect of irrigation water salinity on the relative abundance of mono- and di-glucosylated anthocyanins in fruit arils of 'Wonderful' and 'SP-2' pomegranate accessions.

Soluble Phenolics Content

Arils' soluble phenolics content and response to salinity were similar in both pomegranate accessions (Figure 6). Raising the irrigation water salinity from EC 1.2 to EC 6 was associated with an increase in phenolics concentration; further salinization to EC 9 led to a decline in the phenolics content.

In summary, arils' anthocyanin accumulation decreased with the increase in irrigation water salinity in an accession dependent manner. 'Wonderful' exhibited a high sensitivity whereas 'SP-2" was affected only at the extreme salinity level. Salinity also markedly affected the pigment composition in 'Wonderful' arils. The content of soluble phenolics in the arils of both accessions was comparable and increased with salinity, reaching a maximal value at EC 6.

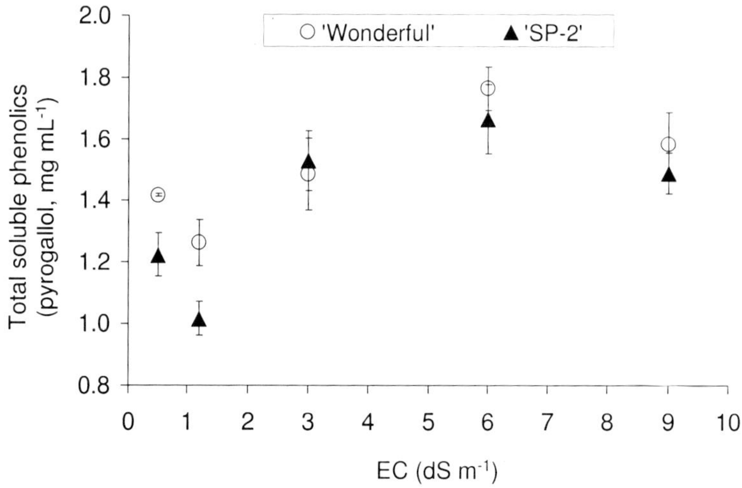

Figure 6. Effect of irrigation water salinity on arils' soluble phenolics content in fruit of 'Wonderful' and 'SP-2' pomegranate accessions.

Discussion

Our results demonstrate that preharvest air temperature and irrigation water salinity markedly affect the two commercially most important traits of the pomegranate fruit, the internal red color intensity and the health potential, through their effects on the anthocyanins (*12–14*) and soluble phenolics (*7–11*) in the fruit arils. The current report gives the first account on salinity impact on the anthocyanins and soluble phenolics in the pomegranate fruit.

Anthocyanins

The red color intensity as well as anthocyanin content and composition in the pomegranate fruit arils vary with season (*20, 31*). Arils' anthocyanin analyses on diverse accessions were combined to present whole year round pattern of changes in relation to the average daily minimal (7-25°C) and maximal (16-40°C) temperatures on harvest date. The temperatures during harvest imply on the preharvest temperature regime. Arils' anthocyanin content was accession dependent but a common negative effect of temperature was established. Anthocyanin accumulation was the highest under cool temperatures, declined with temperature elevation and was remarkably low at hot temperatures. The results support our earlier findings on the negative correlation between arils' red color intensity and the amount of accumulated heat units during pomegranate fruit maturation and ripening (*31*). Taken together, the results point at the temperature as the major cause for seasonal variations in the pomegranate arils' anthocyanin accumulation and, consequently, the fruit internal red color intensity. The mode of response of the pigment accumulation to preharvest temperatures is consistent with the findings in other plant systems (*28–30*), and probably reflects temperature effects on both synthesis and degradation rates. The proportion of diglucosylated pigments positively related to the temperature, and in fruit that ripened under hot temperatures virtually all the anthocyanins were diglucosylated. The latter observation may be explained by the stabilizing effect attributed to anthocyanin glucosylation (*40*, *41*).

Salinity effect on the pomegranate fruit quality was not studied previously. Growing the pomegranate cultivar 'Mollar' in low salt, organic material rich soil was associated with a higher arils' anthocyanin content compared to poor salty soil (*12*). However, no differentiation between the contributions of soil type, salinity, nutrients, and climate was made. Similarly, our previous report on the geographical variations in the pomegranate fruit color and anthocyanin content (*32*) did not discern between the contributions of climate, irrigation water salinity and soil parameters. Arils' anthocyanins were studied under a wide range of salinities, from sweet (EC 0.5) to highly saline (EC 9) water. Increased salinity had a negative effect on anthocyanin accumulation in the pomegranate fruit arils; however, the susceptibility was accession dependent. 'Wonderful' exhibited a high sensitivity to salinity and the anthocyanin content markedly declined already at the lower salinity levels (> EC 0.5), whereas a reduction in anthocyanin content in 'SP-2' arils was only measured at the extreme salinity of EC 9. Salinity also markedly affected anthocyanin composition in 'Wonderful' but not in 'SP-2' arils. The disparity in salinity response between the two accessions may relate to the differences in harvest dates, 'Wonderful' being late-ripening and 'SP-2' early-ripening. We have reported earlier that the effect of differences in environmental conditions on pomegranate fruit parameters were smaller in early ripening compared to mid-season and late ripening accessions (*32*). Alternatively, the difference may reflect the diverse genetic background of the two accessions. Genotype dependent salinity effect was recently reported for anthocyanin accumulation in tomato fruit (*42*).

Taken together, preharvest cool temperatures and low irrigation water salinity are optimal to produce pomegranate fruit with intense red colored arils. Choice of early ripening accessions may alleviate the negative effects of both abiotic factors.

Soluble Phenolics

Phenolics content was measured by the Folin-Ciocalteu method, an electron transfer based assay that measures reducing activity (*43*); thus the reported values for phenolics concentrations represent antioxidant capacities as well. Comparable contents of soluble phenolics were measured in fruit arils of the different accessions. The level of soluble phenolics was stable most of the year but increased by approximately 30% in fruit that ripened during the coolest temperature regime. The impact of temperature on phenolics content in fruit seems to depend on crop and fruit developmental stage. Increasing the preharvest temperature was associated with elevation in the phenolics content in strawberry (*44*) and cranberry (*45*) fruit. Proanthocyanidin content in grapes increased with the increase in heat summation between fruit set and veraison (*46*). However, winter grape berries were significantly richer in phenolics compared to summer berries (*47*), similarly to our findings on pomegranate fruit arils.

Salinity effect on the level of total soluble phenolics in the fruit arils was similar in 'Wonderful' and 'SP-2' accessions. A steep positive effect was measured as the salinity was raised above EC 1.2, reaching a 40-60% increment at EC 6. Further salinization to the extreme salinity of EC 9 was associated with a decline in the phenolics content. Our results are consistent with numerous reports on the increase in phenolics content in plant organs in response to elevated salinity; a function in the plant defense mechanism against salinity induced oxidative stress was proposed (*24*, *27*). Virgin olive oil was considerably richer in phenolics when irrigation was at EC 7.5 compared to EC 1.2 (*48*). Salinity had a positive effect on the phenolics content in red pepper (*49*) and hydroponically grown tomato fruit (*50*, *51*). However, seasonal effects were reported for greenhouse grown tomato; in response to elevated salinity the phenolics content increased in autumn and decreased in spring fruit (*52*).

In summary, the studies indicate that preharvest cool temperatures are associated with enhanced accumulation of anthocyanins and soluble phenolics in the pomegranate fruit arils, and are thus favorable in producing pomegranate fruit with strong internal red color and high health-promoting value. Irrigation with moderately saline water can further raise the health potential of the fruit. However, a concurrent weakening of color intensity may occur, depending on accession. Salinity also affects pomegranate fruit yield in an accession dependent manner (Lazarovitch et al., unpublished results). Choice of the appropriate accession can facilitate salinity utilization to produce acceptable yield of internally red pomegranate fruit with improved health value.

Due to the substantial expansion of pomegranate production to new locations (*6*) and the global trends of warming and water quality deterioration, pomegranate orchards are, and will be more so in the future, exposed to higher temperatures and irrigation water salinities compared to the traditional growing conditions. Our findings can benefit the ongoing efforts to enhance the fruit quality in face of these abiotic factors by breeding and agricultural practices. The information obtained in this study complements the recently acquired knowledge on the molecular biology of anthocyanin biosynthesis pathways in pomegranates ((*53*), Holland et al., unpublished results) and may contribute to the production of cultivars with improved color tolerance to temperature and salinity. Moreover, cultivation approaches that influence the harvest date and include climate and irrigation water salinity management during fruit development and ripening will assist growers in meeting market demands for fruit with intense internal red color and high health-promoting value.

Acknowledgments

The first author, HBN, thanks the Vegetable and Fruit Improvement Center, Texas A&M University, for providing an opportunity to conduct studies during her sabbatical leave. We thank the Jewish National Fund (KKL), the JCA Charitable Foundation and Israel Ministry of Agriculture for financing this study.

References

1. *Pomegranates: Ancient Roots to Modern Medicine*, 1st edition; Seeram, N. P., Schulman, R. N., Heber, D., Eds.; CRC Press, Taylor & Francis Group: Boca Raton, FL, 2006.
2. Bell, C.; Hawthorne, S. Ellagic acid, pomegranate and prostate cancer: A mini review. *J. Pharm. Pharmacol.* **2008**, *60*, 139–144.
3. de Nigris, F; Balestrieri, M. L.; Williams-Ignarro, S.; D'Armiento, F. P.; Fiorito, C.; Ignarro, L. J.; Napoli, C. The influence of pomegranate fruit extract in comparison to regular pomegranate juice and seed oil on nitric oxide and arterial function in obese Zucker rats. *Nitric Oxide* **2007**, *17*, 50–54.
4. Davidson, M. H.; Maki, K. C.; Dicklin, M. R.; Feinstein, S. B.; Witchger, M.; Bell, M.; McGuire, D. K.; Provost, J. C.; Liker, H.; Aviram, M. Effects of consumption of pomegranate juice on carotid intima-media thickness in men and women at moderate risk for coronary heart disease. *Am. J. Cardiol.* **2009**, *104*, 936–942.
5. Seeram, N. P.; Zhang, Y.; Heber, D. Commercialization of Pomegranates: Fresh Fruit, Beverages, and Botanical Extracts. In *Pomegranates: Ancient Roots to Modern Medicine*, 1st edition; Seeram, N. P., Schulman, R. N., Heber, D., Eds.; CRC Press, Taylor & Francis Group: Boca Raton , FL, 2006; pp 187−198.
6. Holland, D.; Hatib, K.; Bar-Ya'akov, I. Pomegranate: Botany, Horticulture, Breeding. *Hortic. Rev.* **2009**, *35*, 127–191.
7. Gil, M. I.; Tomas-Barberan, F. A.; Hess-Pierce, B.; Holcroft, D. M.; Kader, A. A. Antioxidant activity of pomegranate juice and its relationship with phenolic composition and processing. *J. Agric. Food Chem.* **2000**, *48*, 4581–4589.
8. Murthy, K. N. C.; Jayaprakasha, G. K.; Singh, R. P. Studies on antioxidant activity of pomegranate (*Punica granatum*) peel extract using in vivo models. *J. Agric. Food Chem.* **2002**, *50*, 4791–4795.
9. Seeram, N. P.; Lee, R.; Heber, D. Bioavailability of ellagic acid in human plasma after consumption of ellagitannins from pomegranate (*Punica granatum* L.) juice. *Clin. Chim. Acta* **2004**, *348*, 63–68.
10. Seeram, N. P.; Adams, L. S.; Henning, S. M.; Niu, Y.; Zhang, Y.; Nair, M. G.; Heber, D. In vitro antiproliferative, apoptotic and antioxidant activities of punicalagin, ellagic acid and total pomegranate tannin extract are enhanced in combination with other polyphenols as found in pomegranate juice. *J. Nutr. Biochem.* **2005**, *16*, 360–367.
11. Seeram, N.; Lee, R.; Hardy, M.; Heber, D. Rapid large scale purification of ellagitannins from pomegranate husk, a by-product of the commercial juice industry. *Sep. Purif. Technol.* **2005**, *41*, 49–55.
12. Gil, M. I.; Garcia-Viguera, C.; Artes, F.; Tomas-Barberan, F. A. Changes in pomegranate juice pigmentation during ripening. *J. Sci. Food Agric.* **1995**, *68*, 77–81.
13. Gil, M. I.; Cherif, J.; Ayed, N.; Artes, F.; Tomas-Barberan, F. A. Influence of cultivar, maturity stage and geographical location on the juice pigmentation

of Tunisian pomegranates. *Z. Lebensm. Unters. Forsch.* **1995**, *201*, 361–364.

14. Hernandez, F.; Melgarejo, P.; Tomas-Barberan, F. A.; Artes, F. Evolution of juice anthocyanins during ripening of new selected pomegranate (*Punica granatum*) clones. *Eur. Food Res. Technol.* **1999**, *210*, 39–42.
15. Fischer, U. A.; Carle, R.; Kammerer, D. R. Identification and quantification of phenolic compounds from pomegranate (*Punica granatum* L.) peel, mesocarp, aril and differently produced juices by HPLC-DAD–ESI/MS^n. *Food Chem.* **2011**, *127*, 807–821.
16. Du, C. T.; Wang, P. L.; Francis, F. J. Anthocyanins of pomegranate, *Punica granatum*. *J. Food Sci.* **1975**, *2*, 417–418.
17. Noda, Y.; Kaneyuki, T.; Mori, A.; Packer, L. Antioxidant activities of pomegranate fruit extract and its anthocyanidins: Delphinidin, cyanidin, and pelargonidin. *J. Agric. Food Chem.* **2002**, *50*, 166–171.
18. Mavlyanov, S. M.; Islambekov, S. Y.; Karimdzhanov, A. K.; Ismailov, A. I. Polyphenols of the fruits of some varieties of pomegranate growing in Uzbekistan. *Chem. Nat. Compd.* **1997**, *33*, 98–99.
19. Alighourchi, H.; Barzegar, M.; Abbasi, S. Anthocyanins characterization of 15 Iranian pomegranate (*Punica granatum* L.) varieties and their variation after cold storage and pasteurization. *Eur. Food Res. Technol.* **2008**, *227*, 881–887.
20. Borochov-Neori, H.; Judeinstein, S.; Harari, M.; Bar-Ya'akov, I.; Patil, B. S.; Lurie, S.; Holland, D. Climate effects on anthocyanin accumulation and composition in the pomegranate (*Punica granatum* L.) fruit arils. *J. Agric. Food Chem.* **2011**, *59*, 5325–5334.
21. Chalker-Scott, L.; Fuchigami, L. H. The Role of Phenolic Compounds in Plant Stress Responses. In *Low Temperature Stress Physiology in Crops*; Li, P. H., Ed.; CRC Press: Boca Raton, FL, 1989; pp 67–79.
22. Chalker-Scott, L. Environmental significance of anthocyanins in plant stress responses. *Photochem. Photobiol.* **1999**, *70*, 1–9.
23. Chalker-Scott, L. Do Anthocyanins Function as Osmoregulators in Leaf Tissues? In *Why Leaves Turn Red: Anthocyanins in Vegetative Organs. Advances in Botanical Research.* Gould, K. S., Lee D. W., Eds.; Academic Press: London, 2002; pp 103–106.
24. Gould K. S.; Lister C. Flavonoid Functions in Plants. In *Flavonoids: Chemistry, Biochemistry, and Applications*; Andersen, Ø. M., Markham, K. R.., Eds.; CRC Press: Boca Raton, FL, 2006; pp 397–441.
25. Hatier, J.-H. B.; Gould, K. S. Anthocyanin Function in Vegetative Organs. In *Anthocyanins: Biosynthesis, Functions, and Applications*; Gould, K., Davies, K. M., Winefield, C. , Eds.; Springer Science+Business Media, LLC: New York, 2009; pp 1–20.
26. Steyn, W. J. Prevelance and Functions of Anthocyanins in Fruits. In *Anthocyanins:Biosynthesis, Functions, and Applications*; Gould, K., Davies, K. M., Winefield, C. , Eds.; Springer Science+Business Media, LLC: New York, 2009; pp 85–106.
27. Di Ferdinando, M.; Brunetti, C.; Fini, A.; Tattini, M. Flavonoids as Antioxidants in Plants under Abiotic Stresses. In *Abiotic Stress Responses*

in Plants: Metabolism, Productivity and Sustainability; Ahmad, P., Prasad M. N. V., Eds.; Springer Science+Business Media, LLC: New York, 2012; pp 159–180.

28. Oren-Shamir, M. Does anthocyanin degradation play a significant role in determining pigment concentration in plants? *Plant Science* **2009**, *177*, 310–316.
29. Tarara, M. J.; Lee, J.; Spayd, S. E.; Scagel, C. F. Berry temperature and solar radiation alter acylation, proportion and concentration of anthocyanins in Merlot grapes. *Am. J. Enol. Vitic.* **2008**, *59*, 235–247.
30. Mori, K.; Goto-Yamamoto, N.; Kitayama, M.; Hashizume, K. Loss of anthocyanins in red-wine grapes under high temperature. *J. Exp. Bot.* **2007**, *58*, 1935–1945.
31. Borochov-Neori, H.; Judeinstein, S.; Tripler, E.; Harari, M.; Greenberg, A.; Shomer, I.; Holland, D. Seasonal and cultivar variations in antioxidant and sensory quality of pomegranate *(Punica granatum L.)* fruit. *J. Food Compos. Anal.* **2009**, *22*, 189–195.
32. Schwartz, E.; Zulker, R.; Glazer, I.; Bar-Ya'akov, I.; Wiesman, Z.; Tripler, E.; Bar-Ilan, I.; Fromm, H.; Borochov-Neori, H.; Holland, D.; Amir, R. Environmental conditions affect the color, taste and antioxidant capacity of 11 pomegranate accessions fruits. *J. Agric. Food Chem.* **2009**, *57*, 9197–9209.
33. Maas, E. V. Testing Crops for Salinity Tolerance. In *Proc. Workshop on Adaptation of Plants to Soil Stresses*, August 1–4, 1993; Maranville, J. W., Baligar, B. V., Duncan, R. R., Yohe, J. M., Eds.; INTSORMIL. Pub. No. 94-2. University of Nebraska, Lincoln, NE, 1993; pp 234–247.
34. Allen, R. G., Pereira, L. S., Raes, D., Smith, M. *Crop Evapotranspiration: Guidelines for Computing Crop Water Requirements*; FAO Irrig. Drain. Paper 56; UNFAO, Rome, Italy, 1998.
35. Naeini, M. R.; Khoshgoftarmanesh, A. H.; Lessani, H.; Fallahi, E. Effects of sodium chloride-induced salinity on mineral nutrients and soluble sugars in three commercial cultivars of pomegranate. *J. Plant Nutr.* **2004**, *27*, 1319–1326.
36. Naeini, M. R.; Khoshgoftarmanesh, A. H.; Fallahi, E. Partitioning of chlorine, sodium, and potassium and shoot growth of three pomegranate cultivars under different levels of salinity. *J. Plant Nutr.* **2006**, *29*, 1835–1843.
37. Okhovatian-Ardakani, A. R.; Mehrabanian, M.; Dehghani, F.; Akbarzadeh, A. Salt tolerance evaluation and relative comparison in cuttings of different pomegranate cultivars. *Plant Soil Environ.* **2010**, *56*, 176–185.
38. Bhantana, P.; Lazarovitch, N. Evapotranspiration, crop coefficient and growth of two young pomegranate (*Punica granatum* L.) varieties under salt stress. *Agric. Water Manage.* **2010**, *97*, 715–722.
39. Singleton, V. L.; Rossi, J. A. Colorimetry of total phenolics with phosphomolybdic- phosphotungstic acid reagents. *Am. J. Enol. Vitic.* **1965**, *16*, 144–158.
40. Welch, C. R.; Wu, Q.; Simon, J. E. Recent advances in anthocyanin analysis and characterization. *Curr. Anal. Chem.* **2008**, *4*, 75–101.

41. Yonekura-Sakakibara, K.; Nakayama, T.; Yamazaki, M.; Saito, K. Modification and Stabilization of Anthocyanins. In *Anthocyanins: Biosynthesis, Functions, and Applications*; Winefield, C., Davies, K. Gould, K., Eds.; Springer Science+Business Media, LLC.: New York, 2009; pp169–190.
42. Borghesi, E.; González-Miret, M. L.; Escudero-Gilete, M. L.; Malorgio, F.; Heredia, F. J.; Meléndez-Martínez, A. J. Effects of salinity stress on carotenoids, anthocyanins, and color of diverse tomato genotypes. *J. Agric. Food Chem.* **2011**, *59*, 11676–11682.
43. Prior, R. L.; Wu, X.; Schaich, K. Standardized methods for the determination of antioxidant capacity and phenolics in foods and dietary supplements. *J. Agric. Food Chem.* **2005**, *53*, 4290–4302.
44. Wang, S. Y.; Zheng, W. Effect of plant growth temperature on antioxidant capacity in strawberry. *J. Agric. Food Chem.* **2001**, *49*, 4977–4982.
45. Vanden Heuvel, J. E.; Autio, W. R. Early-season air temperature affects phenolic production in 'early black' cranberry fruit. *HortScience* **2008**, *43*, 1737–1741.
46. Pastor del Rio, J. L.; Kennedy, J. A. Development of proanthocyanidins in *Vitis vinifera* L. cv. Pinot noir grapes and extraction into wine. *Amer. J. Enol. Vitic.* **2006**, *57*, 125–132.
47. Xu, C.; Zhang, Y.; Zhu, L.; Huang, Y.; Lu, J. Influence of growing season on phenolic compounds and antioxidant properties of grape berries from vines grown in subtropical climate. *J. Agric. Food Chem.* **2011**, *59*, 1078–86.
48. Ben Ahmed, C.; Ben Rouina, B.; Sensoy, S.; Boukhriss, M. Saline water irrigation effects on fruit development, quality, and phenolic composition of virgin olive oils, CV. Chemlali. *J. Agric. Food Chem.* **2009**, *57*, 2803–2811.
49. Navarro, J. M.; Flores, P.; Garrido, C.; Martinez, V. Changes in the contents of antioxidant compounds in pepper fruits at different ripening stages, as affected by salinity. *Food Chem.* **2006**, *96*, 66–73.
50. de Pascale, S.; Maggio, A.; Fogliano, V.; Ambrosino, P.; Ritieni, A. Irrigation with saline water improves carotenoids content and antioxidant activity of tomato. *J. Hortic. Sci. Biotechnol.* **2001**, *76*, 447–453.
51. Krauss, S.; Schnitzler, W. H.; Grassmann, J.; Woitke, M. The influence of different electrical conductivity values in a simplified recirculating soilless system on inner and outer fruit quality characteristics of tomato. *J. Agric. Food Chem.* **2006**, *54*, 441–448.
52. Incerti, A.; Navari-Izzo, F.; Pardossi, A.; Izzo, R. Seasonal variations in polyphenols and lipoic acid in fruits of tomato irrigated with sea water. *J. Sci. Food Agric.* **2009**, *89*, 1326–1331.
53. Ben-Simhon, Z.; Judeinstein, S.; Nadler-Hassar, T.; Trainin, T.; Bar-Ya'akov, I.; Borochov-Neori, H.; Holland, D. A pomegranate (*Punica granatum* L.) WD40-repeat gene is a functional homologue of Arabidopsis TTG1 and is involved in the regulation of anthocyanin biosynthesis during pomegranate fruit development. *Planta* **2011**, *234*, 865–881.

Chapter 4

Identification of Volatiles from Kumquats and Their Biological Activities

G. K. Jayaprakasha,* C. Kranthi Kumar, Priyanka Chaudhary, K. N. Chidambara Murthy, and Bhimanagouda S. Patil

Vegetable and Fruit Improvement Center, Department of Horticultural Sciences, Texas A&M University, College Station, Texas 77845-2119
***E-mail: gjayaprakasha@ag.tamu.edu.**

Citrus fruits have unique volatile compounds including complex flavoring components. The genus *Fortunella* is taxonomically close to the genus Citrus. Kumquats are a group of small fruit-bearing trees in the flowering plant family Rutaceae either forming the genus *Fortunella*. Generally, these fruits have been used in traditional herbal medicine especially for colds and coughs. Citrus volatiles including kumquats are important targets for studies in chemotaxonomy, biochemical ecology antimicrobial and anticancer activities. The complexity of volatiles often limits to conduct research with target compounds. The advent of gas chromatography in the 1950s allowed researchers to separate volatiles and tentatively identify the major compounds. Later, mass spectral and retention index (RI) of each volatile compounds gives more authenticity for identification using gas chromatography - mass spectrometry (GC-MS). More than 250 volatiles from kumquats and citrus were tabulated along with their Kovats indices using different polarity columns. The formation of volatiles, analytical, separation methods, antioxidant properties and antimicrobial activities of volatile components were also discussed in this chapter.

© 2013 American Chemical Society

Introduction

Kumquats, often known as the little gold gems of citrus family, are specialty crop from Rutaceae family which are related to Citrus spp. Kumquats belong to *Fortunella* genus, which was coined after Robert Fortune (*1*), who introduced kumquat to Europe. They are used for fresh consumption as well as used in marmalades, candied form, and fruit salad. Kumquats are good source of calcium, potassium, vitamin A and vitamin C. In addition, kumquats are also a source of health beneficial secondary metabolites including flavonoids, terpenoids and carotenoids (*2*, *3*). Kumquat essential oil plays an important role in the flavor and aroma of the fruit. Unlike other citrus fruits, kumquats are consumed along with their peel, which provides optimum health benefits obtained from the volatile components and polymethoxy flavones.

Origin and Classification

Citrus fruits include wide range of fruits belonging to six different genera's of subfamily Aurantioideae and family Rutaceae (*4*, *5*). The classification of citrus family is presented in Figure 1. Kumquat belongs to *Fortunella* genus, a native to China and contains six species (*6*, *7*).

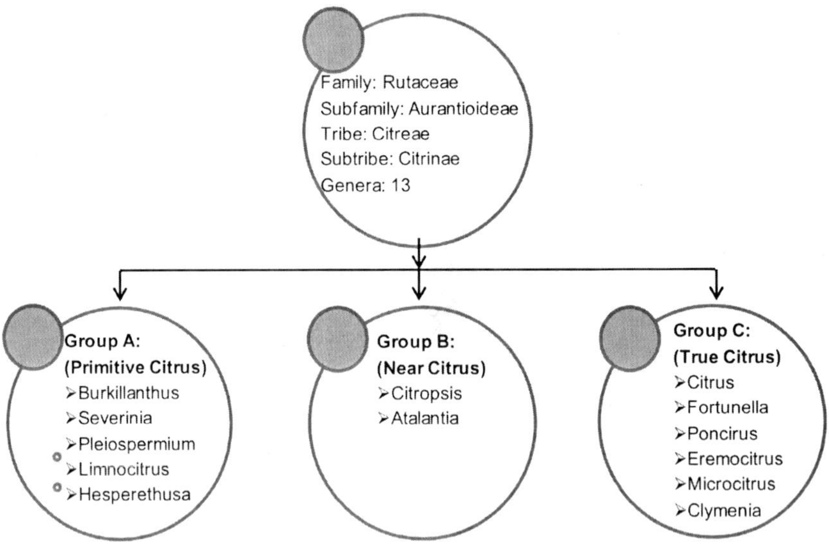

Figure 1. Classification of subfamily Aurantioideae (4, 5).

Kumquats are usually distributed in subtropical temperate regions, except for Malayan kumquat which is distributed in tropical region of Malaysian peninsula and Southern China (*5*, *8*). Kumquat fruits are smaller in size and round or oval in shape. The fruits usually have acidic flesh and sweet peel and are eaten along with the peel (*2*, *9*). They are used in traditional medicine and essential oil production. Kumquats are mainly cultivated in China, Japan, Philippines, Korea, Florida, California and eastern Mediterranean region (*6*, *10*). Different varieties of kumquats and their common names and species are presented in Figure 2.

Scientific name	**Common name**
F. japonica Swing.	Marumi or round kumquat
F. margarita Swing.	Nagami or oval kumquat
F. crassifolia Swing.	Meiwa kumquat
F. hindsii Swing.	Hong Kong wild kumquat
F. polyandra Tan.	Malayan kumquat
F. obovata Tan.	Chagshou kumquat

Nagami kumquat
(*Fortunella margarita*)

Marumi kumquat
(*Fortunella japonica*)

Meiwa kumquat
(*Fortunella crassifolia*)

Malayan kumquat
(*Fortunella polyandra*)

Chagshou kumquat
(*Fortunella obovata*)

Hong Kong kumquat
(*Fortunella hindsii*)

Figure 2. Different species of genus Fortunella and their common names.

In addition, intergeneric hybrids of kumquats have been produced, namely limequat a hybrid between kumquat and lime and citrangequat which is an hybrid between kumquat and citrange (*1*).

Formation of Volatiles in Citrus

Citrus fruits are widely used both as food and as a source of volatile oils. Recently the use of volatile oils has become widespread in foods, drinks, cosmetics, and complementary medicine, especially in aromatherapy (*11*). Citrus juices are the most consumed fruit juices because they combine desirable flavor, appealing color, and health benefits. Volatiles from citrus as well as kumquats are primarily localized in photosynthetically active tissues such as leaves, fruits barring flowers. These volatile oils are stored in specialized oil glands (Figure 3) in the flavedo and oil bodies of juice sacs. The studies conducted in the earlier part of the 20th century have shown development of the oil bearing lumens as schizolysigenous.

Later studies conducted in the 1960s contradicted the previous studies suggesting that the development of lumens is lysigenous (*12*). Due to these confounding results, consensus was not reached until 1976 when Thomson et al., clearly demonstrated schizolysigenous formation of early stage secretory cavities (*13*). The experiments conducted on developmental physiology of secretory cavities have confirmed the results of the 1976 experiment (*12*, *14*).

About 90% of citrus species constitutes of d-limonene as a major volatile principle. D-Limonene is a monoterpene (C_{10}) molecule primarily made up of two isoprene (IPP) units. The biosynthesis of IPP occurs through two separate pathways in compartments of endoplasmic reticulum (mevalonic acid, MVA) and plastids (2-C-methyl-D- erythritol 2,4-cyclodiphosphate, MEP). In cytoplasmic IPP production, three molecules of acetyl-CoA are condensed to IPP with a MVA intermediate. During this condensation, three ATPs and two NADPH molecules are utilized for phosphorylation. However, in plastids, IPP is a produced only after a series of reactions starting with pyruvate (*15*) . In *C.unshiu*, the fruits expressed high levels of terpene synthase in peels at the early stages of the fruit development. Spatio-temporal data mining studies of volatile compounds suggest that these compounds are highly regulated at the stage of transcription. Furthermore, it was also observed that heat and light stress would increase and strongly induce the biosynthesis of volatile terpenoids. Several terpene synthase genes are identified and elucidated using a classical approach called reverse genetics (*16*).

The biosynthesis of volatiles is mainly dependent on the activity of the key enzymes as well as the availability of their substrates. Identification and sequencing the genes responsible for the production of these enzymes and the information related to their expression levels at various fruit developmental stages are critical for engineering these genes for efficient production using alternate hosts such as microbes (*17*) . Several terpene synthases were found to have homologous origin in various fruits and vegetables. Previous studies in lemon showed the sequenced genes for different monoterpene synthases such as γ-terpinene synthase (ClγTS), limonene synthases (Cl(+)LIMS1 and Cl(+)LIMS2), and β-pinene synthase (Cl(-)*β*PINS) which catalyzed the biosynthesis of χ–terpinene, d-limonene and *β*-pinene respectively (*18*, *19*).

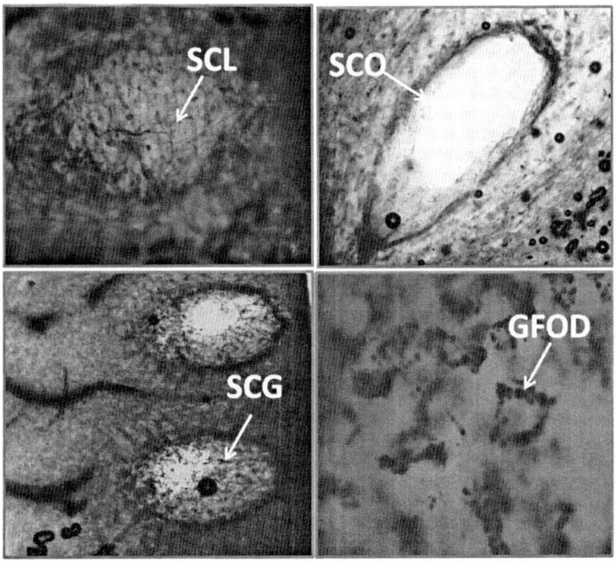

Figure 3. The light micrographic images of secretory cells from lemon (SCL) peel- 5× objective, orange (SCO) peel-5× objective and grapefruit (SCG) peel oil- 2.5× objective captured after staining the sections of flavedo with sudan 4. The grapefruit oil droplets (GFOD)-10× objective were stained red in the intercellular secretory vessels. The stained hand sections were imaged using Zeiss Axiophot microscope equipped with a 2.5x, 10x and 20x Plan Neofluoar objectives and a Nikon DXM1200 color camera and the ACT-1 acquisition software.

It seems that sesquiterpenes (C_{15}), triterpenes (C_{30}) and polyterpenes reproduced in endoplasmic reticulum while isoprenes (C_5), monoterpenes (C_{10}), diterpenes (C_{20}) and tetraterpenes (C_{40}) are in chloroplasts. In general, monoterpene synthases are 600-650 amino acids long while sesquiterpenes are 550–580 amino acids long. In Valencia oranges, the transcript for biosynthesis of valencene tends to accumulate towards fruit maturity. Valencene is the terpenoid that produces the characteristic citrus flavor in Valencia oranges. However, grapefruits have a specific grapefruit flavoring terpenoid called nootkatone, considered to be an oxygenation product of valencene (*20*). It is recently reported that pummelo and its hybrids (grapefruits) produce nootkatone as a defensive response against Drosophila larvae and also an attractant to animals for seed dispersal (*21*). Accumulation of valencene and its putative metabolite nootkatone are highly regulated at the later stages of the fruit maturity and ripening. This suggests that the accumulation of these volatiles is primarily regulated by the ethylene levels in the fruit at the time of maturation.

Chemistry of Kumquat Volatiles

Kumquat volatile oil has typical citrus characteristics (*22*). A diethyl ether extract of the peel of kumquat (*F. crassifolia*) revealed limonene, α-pinene, myrcene, and linalool to be the major constituents (*23*). Some aldehydes, common in citrus chemistry, such as hexanal, octanal, decanal, undecanal, and the powerful (2*E*)-hexenal, (2*E*)-2-nonenal, (2*E*)-2-dodecenal, as well as neral and geranial were also identified (Figure 4 -6).

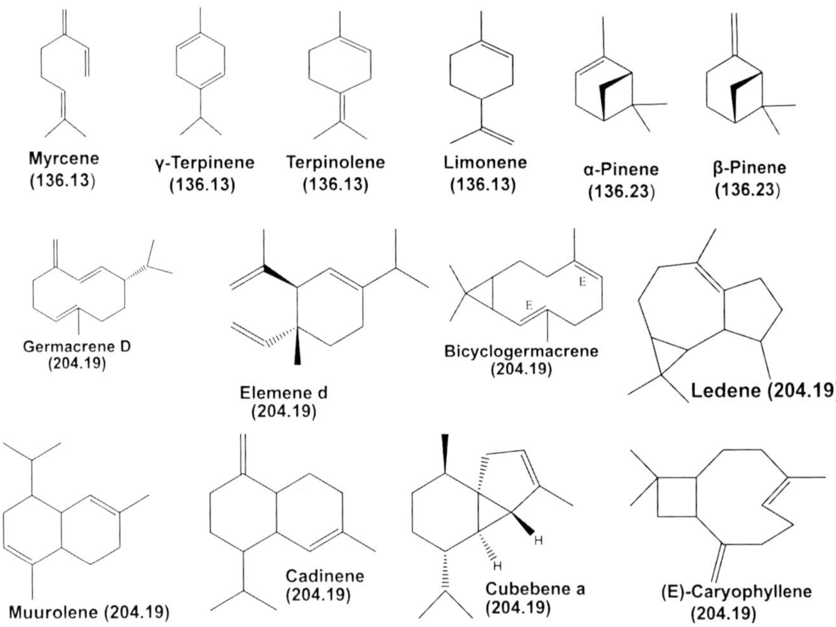

Figure 4. Structures of mono and sesquiterpenes reported from Kumquats. The molecular weight for each compound is mentioned in parentheses.

Recently, the leaf chemical composition of Nagami kumquats, *Fortunella margarita,* was reported with eudesmol (19%), elemol (18.8%) and β-eudesmol (12.4%) as the major compounds (*24*). Eighty-two odor compounds were also identified from *Fortunella japonica* peel oil by GC-MS. Interestingly, d-limonene was found to be a major compound (*22, 25*). More recently, 25 volatile compounds were identified by GC-MS from fresh nagami kumquats (*Fortunella margarita*). The major identified compounds were d-limonene (41.64 %), β-myrcene (16.54 %), linalyl propionate (9.55 %), and germacrene-D (5.93 %) using the Rtx-5 Sil MS column (*26*). Some of the commonly reported compounds from kumquats structures and their exact molecular weights have been presented in Figures 4 - 6.

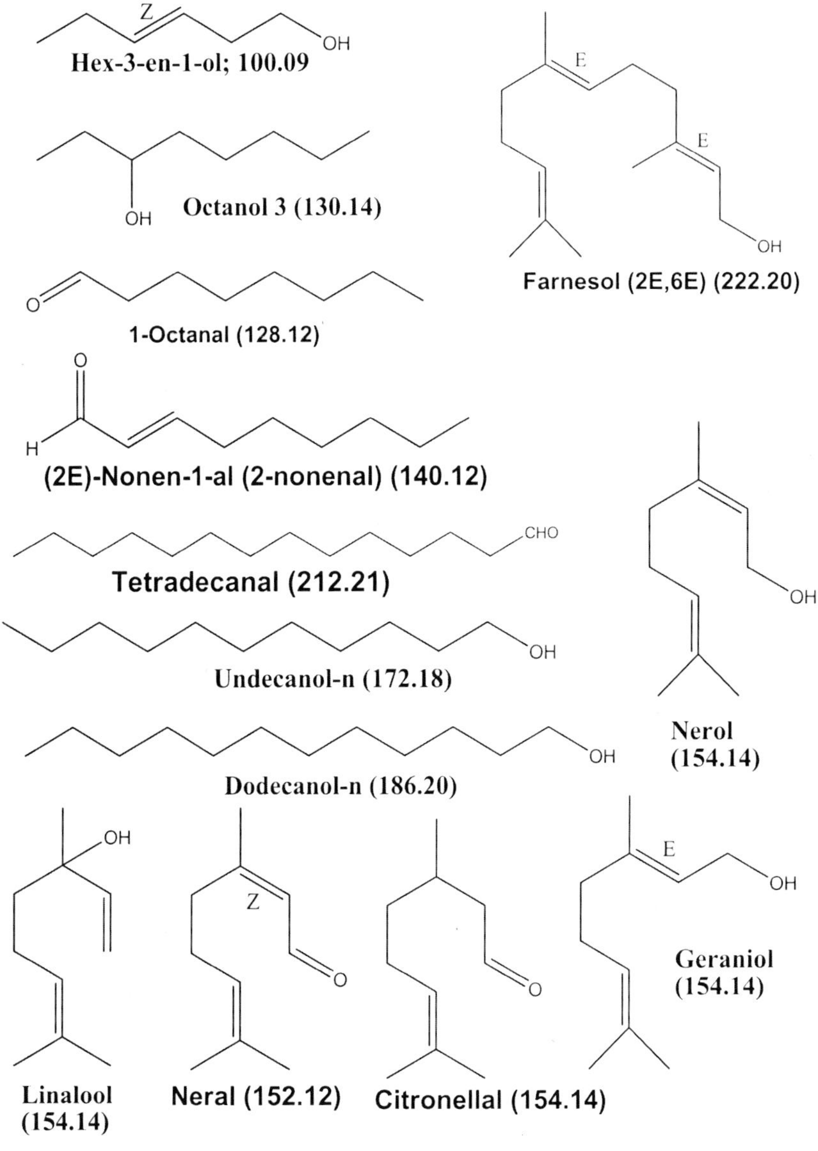

Figure 5. Structures of acyclic oxygenated terpenoids reported from kumquats and citrus. The molecular weight for each compound is mentioned in parentheses.

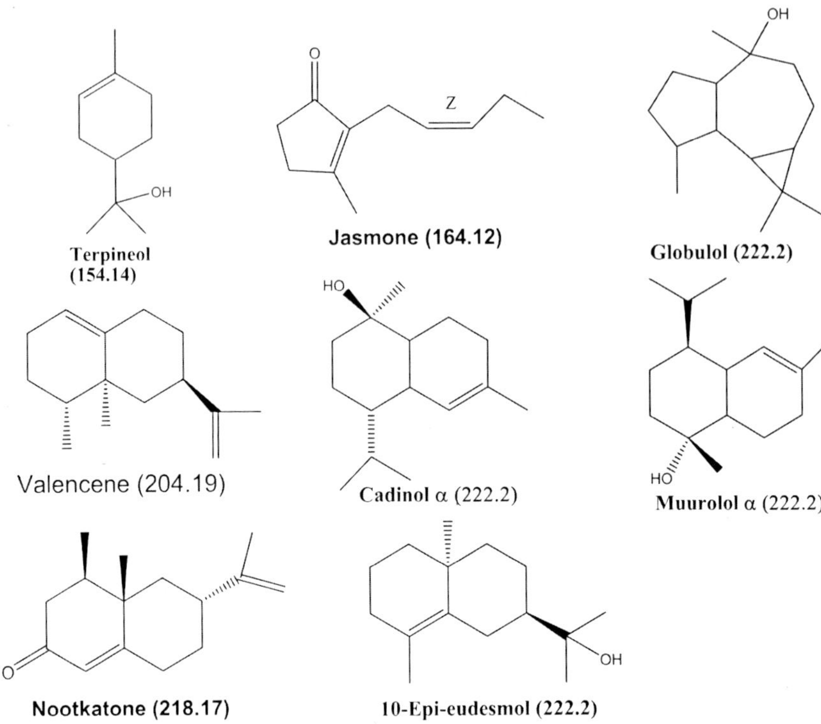

Figure 6. Structures of cyclic oxygenated terpenoids reported from kumquats and citrus. The molecular weight for each compound is mentioned in parentheses.

Compared to other citrus oils, the oil of *F. japonica* contained a larger variety of terpenyl alcohols and esters, such as terpinen-4-ol, *p*-menth-1-en-9-yl acetate, *trans*-*p*-mentha-2,8-dien-1-ol and its acetate, *p*-mentha-1,8-dien-9-yl acetate and its propanoate, as well as higher amounts of sesquiterpenoids (*22*). In general, Kumquat oil contains a higher number of esters and less aldehydes. In addition, nonanal and decanal were almost totally absent. Interestingly, a series of less common terpenyl alcohols and acetates were identified in trace amounts: 2,8-*p*-menthadien-1-ol, *trans*-isopiperitenyl acetate, 1,8(10)-*p*-menthadien-9-yl acetate, 1-*p*-menthen-9-yl acetate, and perillyl acetate (*27*). These compounds may contribute to the fruity-floral character of the peel. In general, citrus oils were classified into two major classes.

a. **Volatile oils:** The citrus volatiles primarily consist of monoterpenes, sesquiterpens including several aldehydes, alcohols and esters. Overall, the volatile fraction of the citrus constitutes to approximately 85-99% of total citrus oils (*28*).
b. **Non-volatile oils:** The non-volatile fraction of the citrus oils consists of fatty acids, hydrocarbons, waxes, sterols, coumarins and polymethoxy flavones. These components account approximately 1-15% of total citrus oils (*28*).

Characteristic Impact Volatile Compounds from Kumquats and Citrus

Until recently, more than 300 volatiles have been identified and reported from citrus including kumquats (*29, 30*) (Table 1). It is a well-established fact that the volatiles that occur in greater concentrations do not necessarily contribute towards greater flavor. Typically, the flavor contributing volatiles are present in lower concentrations compared to those which do not contribute to food aroma (*31*). The volatile compound that provides the characteristic aroma of a food is called as the key odorant or character impact compound. Geranial is the key odorant from lemons (*32–35*). In orange juice, the major odorants are esters (ethyl butanoate) and aldehydes (octanal, nonanal and decanal) (*36*), (*37–39*). The other minor odorants of orange juice include terpene hydrocarbons (α-pinene, β-myrcene, limonene, α-terpinolene), alcohols (1- hexanol, (z)-3-hexan-1-ol, 1- octanol, linalool, geraniol, terpinen-4-ol), ketones (1-octen-3–one, (Z)-octa-1,5-dien-3-one, β-ionone and nootkatone) and oxygen heterocyclic compounds. Andrea et al (*40*) have reported approximately 37 odor active compounds from fresh grapefruit juice. Among these odor imparting compounds *p*-1-menthene-8-thiol, ethyl butanoate, (*Z*)-3-hexenal, 1-hepten-3-one, 4-mercapto-4-methylpentan-2-one and the wine lactone are the major compounds (*40*). The two main factors that contribute to the citrus / kumquats aroma are the chemistry of terpenoids and their concentrations in the matrix. The volatile nature of the compounds is completely dependent on molecular weight of the compounds. If a compound has very low molecular weight, then the compound is highly volatile and vice versa. The polarity of a compound and the type of source matrix would determine the diffusion rate of the volatile. A compound with a lower molecular weight diffuses faster and can be detected at lower temperatures compared to a high molecular weight.

Consumers show affinity towards processed products that have similar aroma components as compare to fresh foods. Thus, identification and purification of these flavoring molecules is essential to understand pollination ecology and plant – animal interactions but it is also critical for a commercial food and perfume industry (*41*). Early studies of citrus volatiles identified d-limonene and α-pinene as the major grapefruit flavor constituent followed by β-caryophyllene, prior to the introduction of gas chromatography in the flavor industry. The quality and composition of the volatiles vary primarily due to environment, maturity stage, cultivar and organs (fruit, flower, branch and leaves) used for extraction. In the grapefruit flower buds, monoterpenes constitute up to 85.7% of total volatiles with sabinene and d-limonene occurs up to 42.2% and 30.8%. As flower matures oxygenated monoterpenes increases (*41*). Approximately 72% of juice volatiles are associated with pulp while 8% are present in the juice (*42*). Furthermore, processing methods and storage conditions greatly influence the odorant composition. In the whole fruits, these volatiles are present in cell organelle compartments where they do not interact with oxidizing and hydrolyzing enzymes. However in fruit juices, these volatiles come in contact with the hydrolyzing enzymes that cause degradation of the original compounds. Thermal processing caused several chemical reactions that would change the original odor of fresh juices (*43*, *44*). According to Baxter et al., 56% of linalool has been converted into degraded compounds such as α-terpineol and 3,7-dimethylact-1-ene-3,7-diol within 20 days after a thermal treatment (*45*). However, this degradation can be avoided by a modified thermal treatment (*32*). Another major factor that alters odor of fresh juice products is the packing material (*46*, *47*).

Sample Preparation Methods for Volatile Components Analysis

Micro-Distillation

For several decades, micro-distillation using Likens and Nickerson apparatus has been used for extraction of volatiles from flavoring materials. As the sample is heated, the volatile compounds along with water vapor are liberated from the sample. The condensation of the vapors would extract the volatiles into pentane (*48*). However, during isolation at higher temperature heat labile compounds may yield several artifacts.

Solid Phase Micro-Extraction

The simultaneous distillation extraction (SDE) has been followed for several decades for the volatiles identification. This method has major limitations including production of artifacts mainly due to thermal decomposition of sugars during extraction at higher temperatures. Pawlisz group first developed, solid phase micro extraction (SPME), which has revolutionized the concept of hyphenation of sample preparation methods and chromatography (GC and HPLC analysis) (*49*). These hyphenated methods have reduced the sample preparation time and disposal costs while increasing the detection limits of the analytes (*50–52*). The SPME device is a modified syringe and the extractions can be performed in the following two different ways (*53*, *54*).

- Direct Immersion (DI) mode can be used for semi-volatiles where the fiber is directly inserted into the liquid sample and the sample is stirred to increase the rate of equilibration. Further, the fiber is retracted into the injection needle and then injected into GC or SPME-HPLC interface and SPME was used to analyze hydrazone-based ligands from orange juice (*55–58*).
- SPME Head Space (HS) technique was adopted for the analysis of volatiles from citrus samples. Recently, SPME-HS analysis was also used for volatile oil analysis from citrus flowers (*41*, *59*, *60*). SPME-HS analysis was considered much cleaner because this method avoids a direct contact with the sample and thereby sustains the durability of the fiber coating (*61*). HS-SPME has no solvent peaks to mask the volatile peaks. Therefore, HS-SPME method is an important role in determining the quantity of the volatile compounds in food samples However, the SPME-HS analysis cannot be used for nonvolatile samples and the unknown volatiles with higher boiling points. The SPME-HS

have shown higher selectivity, reproducibility, precession and accuracy compared to a regular HS analysis (*61*).

Stir Bar Sorptive Extraction

The major disadvantage of SPME roots to its low adsorption capacity which intern causes lower extraction of the compounds that are partially water soluble. Additionally, in SPME technique, acute competition between polydimethylsiloxane (PDMS) fiber, glass wall and the stir bar used for stirring samples (*62*). The stir bar micro extraction (SBME) was a solely developed to have micro sensitivity for the detection of trace volatiles. A commonly PDMS fiber in SPME with a 100 μM length have the capacity adsorption upto 0.5 μL. The lower adsorption capacity of the SPME was overcome with the use of SBME which has 50-250 times more adsorption capacity due to its higher phase ratio (*62*). The SBSE also uses PDMS as sorptive material for extraction similar the SPME but requires a separate desorption chamber which is done thermally or with organic solvent. Despite the huge potential for SBME in citrus volatile research, there is no evidence of using SBSE in identification of trace or unknown citrus volatiles. After desorption, the sample is cryo-focused into a GC column for analysis.

Separation of Volatiles

In general, gas chromatography is widely used in the separation and partial characterization of the volatiles. The molecular mass at which GC operates ranges from 2-1500 atomic mass units. Therefore, the compounds such as permanent gases, volatile (monoterpenes) and partially volatile (atomic mass ≥ 200) compounds can be separated and identified on a GC. Any good volatile separations in GC would primarily dependent on (a) the stationary phase of the column (b) peak capacity. The former aspect of GC column is primarily dependent on the chemical composition of the stationary phase while the later feather is governed by factors such as stationary phase thickness, length and diameter of the column, and analyte-stationary phase interactions (*63*). Column selectivity is very critical for separation of volatiles with very similar properties. For example, *cis* and *trans* isomers can be separated on the stationary phase made of cyclo-dextrin derivatives which forms the inner lining of chiral columns (*64*). Single column (one-dimensional) chromatography analysis has been widely used for many years as a standard separation tool for analyzing volatile compounds in a broad variety of fields including food analysis (*65*, *66*). At present, attention is being paid to avoid laborious sample pre-treatments that can in fact be an important source of errors mainly for complex matrices as e.g., food or food-related matrices. One-dimensional chromatography does not always provide the resolution and separation power required to obtain the best results in terms of identification of analytes in food samples. Multidimensional chromatography has emerged as an interesting alternative to analyze complex samples in a situation in which technological improvements, such as new column technologies, seem to be close to their maximum level. Thus, peak capacity enhancement achievable by multidimensional chromatography is by far higher than the obtained after improving by any mean one-dimensional separations. Multidimensional chromatography allows combination of two or more independent or nearly independent separation steps, increasing significantly the separation power of the corresponding one-dimensional techniques (*67*, *68*).

The following are some detection methods commonly used for analysis of the volatiles.

Identification of Volatiles

GC-FID and GC-O Analysis

Flame ionization detector (FID) is one of the oldest and common detector used in the analysis of volatiles. The use of FID is popular amongst the perfumery and food industry is mainly due to its high sensitivity. Gas chromatography-olfactometry (GC-O) was a widely used detection technique in the food and perfumery industries for identification of certain specific chemicals that impart odor to the sample (*54*, *69*, *70*). A small modification was made into the eluent line, which branches out into a conventional detector and a funnel like glass chamber. The analyst would sniff at the port of the funnel that branched out from the eluent line. At this point the analyst would describe the odor and intensity during the peak elution. The analyst adjusts a dilution apparatus (aroma extract dilution analysis, AEDA) with in GC-O to the highest dilution level to evaluate the odor threshold of a volatile compound.

FT-IR

Infrared spectroscopy is considered as one of the most versatile non-destructive form of detection that is available especially for compounds that have oxygen, hydrogen and nitrogen molecules (*71*). Due to its low sensitivity (slow scanning process), newly developed fourier transform infrared (FT-IR) spectrometry has provided a practical alternative for infrared spectroscopy. The FT-IR is equipped with a beam splitting optical device called interferometer. The interferometer is equipped with two mirrors, one fixed and another moving. The infrared (IR) beam is spit and focused on to the two mirrors. The reflected IR beams from the mirrors are combined to form another wave called interferogram. This interferogram is transmitted through the sample and specific frequencies of IR beam are absorbed by the sample. Finally, the IR beam emitted through the sample passes through the detector and the signal in the detector is digitized in the computer where Fourier transformation occurs.

Natural volatile compounds are commonly identified using gas chromatography (GC) in combination with (1) mass spectroscopy (MS) for spectral matching, (2) flame ionization detection (FID), (3) GC-O and (4) retention index (RI) matching for quantification and identification of peak from diverse unknowns. A single plant volatile oil may contain hundreds of structurally similar terpenoids that produce highly similar mass spectra. Thus, compound matching by mass spectrometry alone is error prone. Matching by comparing retention times with those of standards run on the same system can be impractical, since many plant compounds are not commercially available. Published retention indices are more suitable than retention times for comparison across different chromatographic systems. The combined matching of mass spectra and retention indices can facilitate positive identification of more than 100 compounds from a single plant extract with only a few chromatograms.

Identification of Volatiles by Mass Spectra with Kovats Indices

Retention indices for all the compounds were determined according to the Kovats method (*72*) using *n*-alkanes as standards. Kovats retention indices were calculated using *n*-paraffins (C_6-C_{26}) as references. KI = 100 N + 100 n [log t^1 R (A) − log t^1 R (N)] / [log t^1 R (N + n) − log t^1 R(N)], Where t^1 R(N) and t^1 R (N+ n) are the adjusted retention times of n-paraffin hydrocarbons of carbon number N and (N + n) that are respectively smaller and larger than the adjusted retention time of the unknown t^1 (R). The comprehensive list of volatiles reported from kumquats and citrus has been tabulated in Table 1 along with Kovats indices using non-polar, medium polar and polar column (*40*, *73–78*).

Table 1. Gas Chromatographic Separations of Various Citrus Volatiles in Non-Polar, Polar and Intermediate Polar Columns and Their Respective Kovats Indices

Volatile Compound	*Non-polar column*	*Polar Column*	*Medium polar column*	*Reference*
Acetic acid	610	1450	781	(*40*)
Acetone	-	831	-	(*74*)
Benzaldehyde	-	1573	-	(*74*)
Benzene acetonitrile	-	1982	-	(*74*)
Benzofuranone 3a,4,5,7a-tetrahydro-3,6-dimethyl-2(3H)-	1456	2220	1687	(*40*)
Bergamotene	-	1593	-	(*75*)
Bergamotene α-	1435	1589, 1615	-	(*74, 76*)
Bergamotene α-cis-	1418	-	-	(*77*)
Bisabolene α-	1496	1724	-	(*78*)
Bisabolene β-	1495, 1513	1754, 1770	-	(*74, 76, 77*)
Bisabolene	-	1777	-	(*75*)

Continued on next page.

Table 1. (Continued). Gas Chromatographic Separations of Various Citrus Volatiles in Non-Polar, Polar and Intermediate Polar Columns and Their Respective Kovats Indices

Volatile Compound	*Non-polar column*	*Polar Column*	*Medium polar column*	*Reference*
Bisabolol	-	2224	-	(*75*)
Bisabolol α-	1697	-	-	(*77*)
Butane-2,3-dione	<600	970	692	(*40*)
Butanoic acid	821	1622	996	(*40*)
Cadinene δ-	1511, 1516, 1527	1766, 1746	-	(*76–78*)
Cadinene	-	1764	-	(*75*)
Cadinol τ-	1629	2171	-	(*78*)
Calamenene	-	1879	-	(*74*)
Camphene	952	1083	-	(*75, 79*)
Camphor d-	-	1532	-	(*75*)
Carene 3-	1000	1150	-	(*76*)
Carene d-3-	1012	1166	-	(*74, 79*)
Carene δ-3-	1005	1145, 1162	-	(*75, 78*)
Carvacrol	1303	-	-	(*77*)
Carveol (E)-	-	1857	-	(*75*)
Carveol (Z)-	-	1841	-	(*75*)
Carveol cis-	1237, 1238	-	-	(*77, 79*)
Carveol trans-	1223, 1225	-	-	(*77, 79*)
Carvone	1248, 1251	-	-	(*77, 79*)
Carvone (Z)-	-	1744	-	(*75*)
Carvyl acetate	-	1475	-	(*74*)
Caryophyllene (E)-	1420	1588	-	(*78*)
Caryophyllene oxide	1591, 1595	-	-	(*77, 79*)
Caryophyllene β-	1424, 1427, 1428	1654, 1642, 1609	-	(*74–77, 79*)
Cedrene α-	-	1577	-	(*75*)
Cedrenol	-	2129	-	(*75*)
Cineole 1,4-	1009	1177	-	(*76*)

Continued on next page.

Table 1. (Continued). Gas Chromatographic Separations of Various Citrus Volatiles in Non-Polar, Polar and Intermediate Polar Columns and Their Respective Kovats Indices

Volatile Compound	*Non-polar column*	*Polar Column*	*Medium polar column*	*Reference*
Cineole 1,8-	1037	1232	-	(*74*, *77*)
Cineole	-	1236	-	(*75*)
Citronellal	1131,1155, 1130	1472, 1505, 1473, 1486	-	(*74–79*)
Citronellol	1203, 1207, 1235	1751, 1756	-	(*76–78*)
Citronellol α-	-	1772	-	(*74*)
Citronellol β-	-	1777, 1771	-	(*74*, *75*)
Citronellyl acetate	1335	1645, 1665	-	(*75*, *76*)
Comphene (camphene)	962	1083	-	(*76*)
Copaene α-	-	1528, 1499	-	(*74*, *75*)
Copaene β-	1438	-	-	(*77*)
Cryptone	1195	-	-	(*77*)
Cubebene	-	1550	-	(*75*)
Cubebene α-	-	1468	-	(*75*)
Cubenene β-	1393	-	-	(*77*)
Cumin aldehyde	-	1791	-	(*75*)
Curcumene	-	1804	-	(*74*)
Cymene p-	1013, 1027, 1012, 1028	1284, 1299, 1267, 1282	-	(*74–79*)
Cymenene p-	1092, 1072	1432	-	(*78*, *79*)
Decanal	1183, 1209	1482, 1501, 1505	-	(*36*, *75–77*, *79*)
Decanal green	1204	1493	1290	(*40*)
Decanol	1270, 1276	1729, 1768	-	(*75*, *76*, *79*)
Decenal (E)-2-	1266	1651	-	(*75*, *79*)
Decenal 4,5-epoxy-(E)-2-	1380	2000	1552	(*40*)

Continued on next page.

Table 1. (Continued). Gas Chromatographic Separations of Various Citrus Volatiles in Non-Polar, Polar and Intermediate Polar Columns and Their Respective Kovats Indices

Volatile Compound	*Non-polar column*	*Polar Column*	*Medium polar column*	*Reference*
Decyl acetate	1411	-	-	(*79*)
Dihydro linalool 1,2-	1121	1449	-	(*76*)
Dimethyl anthranilate	-	2142	-	(*74*)
Dimethylstyrene α-p-	-	1471	-	(*74*)
Dodecadienal (E,Z)-2,6-	1452	-	-	(*79*)
Dodecanal	1410	1713	-	(*75, 79*)
Dodecenal (E)-2-	1471	1867	-	(*75, 79*)
Dodecyl acetate	1611	-	-	(*79*)
Elemene d-	1340	1498	-	(*74, 79*)
Elemene β-	1395, 1388	1624, 1586, 1598	-	(*74, 75, 77, 78*)
Elemene δ-	1384	1475	-	(*76*)
Elemol β-	-	2088	-	(*75*)
Epoxy limonene Cis-	1118	-	-	(*76*)
Epoxy ocimene (Z)-	-	1512	-	(*74*)
Ethanol	-	949	-	(*74*)
Ethyl 2-methyl propanoate	751	955	812	(*40*)
Ethyl 3-hydroxy hexanoate	1134	1674	1245	(*40*)
Ethyl acetate	<600	888	<600	(*40*)
Ethyl butanoate	802	1049, 1028	856	(*36, 40*)
Ethyl hexanoate	1002	1246, 1226	1058	(*36, 40*)
Ethyl octanoate	-	1448	-	(*36*)
Ethyl propanoate	714	951	765	(*40*)
Ethyl-2-methyl butanoate	-	1063	-	(*36*)
Eugenol	-	2178	-	(*75*)
Farnesene (E)-β-	1456	-	-	(*79*)
Farnesene (E,E)-α-	1740	1501	-	(*78*)
Farnesene (Z)-β-	1455	-	-	(*77*)
Farnesene α-	-	1732	-	(*75*)

Continued on next page.

Table 1. (Continued). Gas Chromatographic Separations of Various Citrus Volatiles in Non-Polar, Polar and Intermediate Polar Columns and Their Respective Kovats Indices

Volatile Compound	*Non-polar column*	*Polar Column*	*Medium polar column*	*Reference*
Farnesene β-	-	1681, 1650	-	(*74, 75*)
Farnesol	-	2375	-	(*74*)
Farnesol (E,E)-	-	2353	-	(*75*)
Farnesol (Z,E)-	-	2291	-	(*75*)
Furaneol acetate (Furanone 4-hydroxy-2,5-dimethyl-3(2H)-)	1062	2024	1242	(*40*)
Geranial	1248, 1272, 1241,	1735, 1766, 1741, 1738	-	(*74–79*)
Geraniol	1237, 1252, 1232	1789, 1861, 1837, 1849	-	(*74–76, 78, 79*)
Geranyl acetate	1380, 1358	1748, 1760	-	(*75, 78, 79*)
Geranyl formate	-	1709	-	(*75*)
Germacrene	1490	-	-	(*77*)
Germacrene D	1477, 1488, 1480	1714, 1704, 1719	-	(*75, 76, 78, 79*)
Germacrene Bicyclo	1503, 1505	1744	-	(*75, 77, 79*)
Heptadecane	-	1699	-	(*74*)
Heptadecene 8-	-	1725	-	(*74*)
Hepten-3-one 1-	886	1190	954	(*40*)
Heptyl acetate	-	1113	-	(*79*)
Hexanal	-	1100, 1099	-	(*36, 74*)
Hexanol 1-	-	1364	-	(*74*)
Hexenal Z-3	-	1153	-	(*36*)
Hexyl acetate (E)-2-	1014	-	-	(*79*)
Humulene α-	1463, 1453	1660, 1680	-	(*78, 79*)
Indole	-	2507	-	(*74*)
Isopropanol	-	940	-	(*74*)

Continued on next page.

Table 1. (Continued). Gas Chromatographic Separations of Various Citrus Volatiles in Non-Polar, Polar and Intermediate Polar Columns and Their Respective Kovats Indices

Volatile Compound	*Non-polar column*	*Polar Column*	*Medium polar column*	*Reference*
Iso pulegol	1146	1565	-	(*76*)
Jasmone (Z)-	-	1996	-	(*74*)
Limonen-10-ol	1295	-	-	(*79*)
Limonene	1036, 1021, 1033	1219, 1235, 1199, 1230	-	(*36, 74, 75, 77–79*)
Limonene (R)-	1025	1188	1054	(*40*)
Limonene d-	1026	1201	-	(*76*)
Limonene di epoxide	-	2059	-	(*75*)
Limonene oxide (E)-	-	1472	-	(*75*)
Limonene oxide (Z)-	-	1460, 1487	-	(*75*), (*74*)
Limonene oxide cis-	1136, 1138	-	-	(*77, 79*)
Limonene oxide trans-	1140	-	-	(*79*)
Linalool	1083, 1102, 1082	1553, 1561, 1560, 1539, 1552	-	(*36, 74–79*)
Linalool oxide (E)-	-	1480	-	(*75*)
Linalool oxide (Z)-	-	1453	-	(*75*)
Linalyl acetate	1252, 1239	1577, 1549, 1571	-	(*74, 75, 78, 79*)
Me N-methyl anthranilate	1385	2082	-	(*78*)
Menth-1-en-9-ol p-	-	1560	-	(*75*)
Menth-2-en-1-ol cis-p-	1107	1555	-	(*78*)
Menth-2-en-1-ol trans-p-	1123	1618	-	(*78*)
Mentha-1(7),8-dien-2-ol cis-p-	1243	-	-	(*77*)
Mentha-1(7),8-dien-2-ol trans-p-	1190	-	-	(*79*)
Mentha-1,8-dien-3-one (+)-p-	1277	-	-	(*77*)
Mentha-1,8-dien-7-ol p-	1206	-	-	(*77*)

Continued on next page.

Table 1. (Continued). Gas Chromatographic Separations of Various Citrus Volatiles in Non-Polar, Polar and Intermediate Polar Columns and Their Respective Kovats Indices

Volatile Compound	*Non-polar column*	*Polar Column*	*Medium polar column*	*Reference*
Mentha-2,8-dien-1-ol cis-p-	1142	-	-	(*77*)
Mentha-2,8-dien-1-ol trans-p-	1125	-	-	(*79*)
Mentha-2,8-dienol trans-p-	1127	-	-	(*77*)
Menth-2,8-dien-1-ol (E)-p-	-	1637	-	(*75*)
Menthene-8-thiol p-1-	1283	1598	1348	(*40*)
Menthone	-	1475	-	(*75*)
Mercapto-4-methylpentan-2-one 4-	937	1377	1053	(*40*)
Methional	900	1449	1040	(*40*)
Methoxypyrazine, 2-Isopropyl-3-	1092	1427	1143	(*40*)
Methyl acetate	-	841	-	(*74*)
Methyl anthranilate	-	2299	-	(*74*)
Methyl butanoate	-	993	-	(*36*)
Methyl furan 2-	-	910	-	(*74*)
Methyl salicylate	-	1832	-	(*74*)
Methyl-5-hepten-2-one 6-	-	1361	-	(*74*)
Methyl butanoate (S)-ethyl 2-	845	1041	907	(*40*)
Methyl butanoic 2-,3-	875	1660	1030	(*40*)
Methyl butanol 2, 3	739	1211	835	(*40*)
Methyl butyl butyrate 2-	-	1289	-	(*75*)
Methyl phenylacetate	-	1799	-	(*74*)
Methyl thymol	1196	-	-	(*76*)
Muurolene α-	1502, 1506	1730	-	(*76, 77*)
Myrcene	991, 978, 990, 989	1157, 1169, 1154	1020	(*40, 75, 77–79*)
Myrcene α-	-	1183	-	(*74*)
Myrcene β-	982	1176, 1162, 1175	-	(*36, 74, 76*)

Continued on next page.

Table 1. (Continued). Gas Chromatographic Separations of Various Citrus Volatiles in Non-Polar, Polar and Intermediate Polar Columns and Their Respective Kovats Indices

Volatile Compound	*Non-polar column*	*Polar Column*	*Medium polar column*	*Reference*
Neral	1215, 1242, 1213	1705, 1674, 1688	-	(*75, 76, 78, 79*)
Nerol	1208, 1207	1832, 1818, 1853	-	(*74–76, 78*)
Nerolidol (E)-	1565, 1547	2031, 2042	-	(*75, 78, 79*)
Nerolidol (Z)-	-	1995	-	(*75*)
Nerolidolc	-	2054	-	(*74*)
Neryl acetate	1361, 1340	1725, 1730	-	(*75, 78, 79*)
Nonanal	1082, 1107, 1102	1382, 1411, 1383	1193	(*36, 40, 76, 79*)
Nonanal n-	1107	-	-	(*77*)
Nonanoic acid	-	2196	-	(*75*)
Nonanol	1161, 1175	-	-	(*76, 79*)
Nonatriene 4,8-dimethyl-1,3 (E),7-	1115	-	-	(*79*)
Nonenal (E)-2-	1530	1157	1278	(*40, 79*)
Nonenal (Z)-2-	1143	1500	1252	(*40*)
Nonyl acetate	-	1581	-	(*75*)
Nootkatone	1815, 1814, 1826	2515, 2544	2006	(*40, 75, 77, 79*)
Ocimene (E)-	-	1267	-	(*74*)
Ocimene (E)-β-	1048, 1035, 1048	1245, 1258	-	(*75, 77–79*)
Ocimene (Z)-β-	-	1234	-	(*75*)
Ocimene (Z)-	-	1248	-	(*74*)
Ocimene (Z)-β-	1024	1228	-	(*78*)
Ocimene cis-β-	1028	1235	-	(*76*)
Ocimene trans- β-	1035	1252	-	(*76*)
Ocimene Allo	-	1421	-	(*74*)

Continued on next page.

Table 1. (Continued). Gas Chromatographic Separations of Various Citrus Volatiles in Non-Polar, Polar and Intermediate Polar Columns and Their Respective Kovats Indices

Volatile Compound	*Non-polar column*	*Polar Column*	*Medium polar column*	*Reference*
Ocimene Neo-allo-	-	1396	-	*(74)*
Octadecane	-	1805	-	*(75)*
Octadien-3-one 1,5(Z)-	982	1367	1081	*(40)*
Octadiene 2,7-dimethyl-1,6-	-	1312	-	*(75)*
Octanal	1005, 1006	1298	-	*(75, 77, 79)*
Octanal n-	977	1270	-	*(76)*
Octanol	1073	1566	-	*(75, 79)*
Octanol n-	1074, 1073	1507	-	*(76, 77)*
Octatetraene 2,6-Dimethyl-1,3,5,7-	-	1459	-	*(74)*
Octen-3-one 1-	986	1292	1067	*(40)*
Octyl acetate	1212	-	-	*(79)*
Octyl formate	1129	-	-	*(79)*
Patchenol (E)-	1337	-	-	*(77)*
Patchenol (Z)-	1325	-	-	*(77)*
Pentadecane	-	1501	-	*(74)*
Penten-3-one 1-	683	1026	765	*(40)*
Perillyl alcohol	1304, 1306	2003	-	*(75, 77, 79)*
Perilla aldehyde	1284, 1265, 1280	1794	-	*(75–77, 79)*
Perillyl acetate	1438	-	-	*(79)*
Phellandrene α-	994	1009, 1177, 1179	-	*(75, 76, 78, 79)*
Phellandrene β-	1021, 1035	1208, 1235	-	*(75, 77, 78)*
Phenethyl alcohol	-	1950	-	*(74)*
Phenylacetaldehyde	1050	1691, 1639	1175	*(40, 74)*
Phenylformamide n-	-	2232	-	*(74)*
Pinene (R)-R-	929	1010	945	*(40)*

Continued on next page.

Table 1. (Continued). Gas Chromatographic Separations of Various Citrus Volatiles in Non-Polar, Polar and Intermediate Polar Columns and Their Respective Kovats Indices

Volatile Compound	*Non-polar column*	*Polar Column*	*Medium polar column*	*Reference*
Pinene α-	935, 932, 930, 936	1036, 1021, 1030, 1037	-	(*36*, *74–79*)
Pinene β-	981, 973, 971, 981	1125, 1107, 1118, 1109, 1126	-	(*36*, *74–79*)
Pinocarveol cis-	1193	-	-	(*77*)
Piperitol (E)-	-	1693	-	(*75*)
Sabinene	967, 975, 964	1120, 1137, 1119, 1134	-	(*75*, *77–79*) (*74*, *76*)
Sabinene hydrate	-	1485	-	(*74*)
Sabinene hydrate Cis-	1088, 1074	-	-	(*76*)
Sabinene hydrate trans-	1054, 1053	1463, 1458	-	(*76*, *78*)
Santalene β-	1465	-	-	(*77*)
Santalene	-	1656	-	(*75*)
Sativene	1397	-	-	(*77*)
Selinene 7-epi-α-	1532	-	-	(*77*)
Selinene β-	1493	-	-	(*77*)
Sesquiphellandrene	-	1782	-	(*75*)
Sesquiterpenol	1650	-	-	(*76*)
Sinensal α-	1725, 1726	2323, 2340	-	(*75*, *76*, *78*)
Sinensal β-	1673, 1701	2225, 2237	-	(*75*, *77*, *78*)
Spathulenol	1586, 1589	-	-	(*77*, *79*)
Sylvestrene iso-	1012	-	-	(*77*)
Terpinen-4-ol	1168, 1185, 1162	1610, 1595, 1603	-	(*75*, *76*, *78*, *79*)
Terpinene c-	-	1285	-	(*74*)
Terpinene α-	1018, 1009, 1020,	1199, 1179, 1193	-	(*74*, *75*, *77–79*)

Continued on next page.

Table 1. (Continued). Gas Chromatographic Separations of Various Citrus Volatiles in Non-Polar, Polar and Intermediate Polar Columns and Their Respective Kovats Indices

Volatile Compound	*Non-polar column*	*Polar Column*	*Medium polar column*	*Reference*
Terpinene γ-	1051, 1061, 1048, 1061	1248, 1241,1264	-	*(75–79)*
Terpineol trans-b-	1151	-	-	*(79)*
Terpineol α-	1179, 1197, 1172, 1201	1740, 1688, 1703	-	*(75–79)*
Terpineol-4-ol	1186	-	-	*(77)*
Terpinolene	1080, 1089, 1078,	1291, 1278,1295	-	*(75–79)*
Terpinolene α-	-	1307	-	*(74)*
Terpinolene/octanal α-	-	1319	-	*(36)*
Terpinyl acetate α-	1351	-	-	*(79)*
Terpinyl acetate	-	1698	-	*(75)*
Tetradecane	-	1400	-	*(75)*
Tetradecenal (E)-2-	1677	-	-	*(79)*
Thujene α-	924, 922, 927	1025, 1021, 1039	-	*(74, 76–79)*
Thujopsene cis-	1438	-	-	*(77)*
Thymal	1286	2100	-	*(76)*
Thymol	1266, 1295	2208, 2189	-	*(74, 77, 78)*
Thymol methyl ether	-	1621	-	*(74)*
Thymyl methyl oxide	1214	1589	-	*(78)*
Tridecane	-	1301	-	*(75)*
Trimenal (Z)-	1413	-	-	*(77)*
Undecanal	1309	1610	-	*(75, 79)*
Valencene	1500	1714	-	*(75, 77)*
Vanillin	1397	2567	1638	*(40)*

NP: Non-polar column (includes SE-54, BP-1, DB-5, SLB-5, DB-1); MP - medium polar column (OV-1701) and polar columns (DB Wax, FFRAP, BP-20).

Effect of Postharvest Treatments on Volatile Oil Composition

As compared to other *Citrus* species less attention has been given to effect of postharvest treatments on volatile oil composition of *Fortunella* species. Various postharvest treatments are practiced in kumquats to enhance the shelf life and quality of fruits (*80*). Most common treatment is hot water dipping (*80*, *81*) to prevent chilling injury and decay in kumquats. Hot water dipping (HWD, 2 min at 50 °C) had no significant effect on major components of kumquat essential oils with only increase observed in p-mentha-1,5-dien-1-ol (*3*). On other hand, storage period increased the compounds responsible for off-flavors such as p-cymene and p-cymen-8-ol in the control fruits which was not observed in HWD treated fruits (*3*). There is tremendous scope to study the effect of various postharvest and pre-harvest treatments on the volatile composition in kumquats.

Antioxidant Activity

Antioxidants play an important role in quenching the free radicals and oxidants present in the body, maintaining a healthy balance for normal metabolism (*82*). Fruits and vegetables are rich source of antioxidants (*83*). Citrus fruits have received lot of attention towards antioxidant activity and phytochemicals as compared to the minor species including *Fortunella*. Citrus volatile oils contain antioxidant properties due to presence of certain oxygenated terpenes (*84–86*). In a study conducted by Choi et al, 34 different citrus essential oils and 21 authentic compounds were analyzed for their antioxidant activity. While D-Limonene and myrcene showed lower radical scavenging activity, α-terpinene, nootkatone, citronellal, citral, geraniol, γ-terpinene and terpinolene contain higher radical scavenging activity (*84*). Terpenes, such as γ-terpinene and terpinolene, which are considered as important contributors to the radical scavenging activity (*84*), are present in minor quantities in kumquat essential oil (*2*, *9*, *87*). Geraniol has been reported to possess highest radical scavenging activity (*84*).

In addition to the volatile oils, studies were conducted to understand the antioxidant potential of kumquat extracts (*88*). Different fractions of Nagami kumquats (*Fortunella margarita)* were analyzed for their antioxidant activity. Highest diphenyl picryl hydrazyl (DPPH) radical scavenging activity was found in ethyl acetate and acetone extracts while methanol-water (4:1, v/v) extracts had higher oxygen radical absorbance capacity values and antioxidant capacity using phosphomolybdenum method. In another study conducted on Nagami kumquat similar results were observed where EtOAc fractions showed highest antioxidant activity using DPPH assay (*89*). Kumquat peels are good source of polyphenols along with volatile oils (*88*).

Inhibitory Effect on NDMA Formation

N-nitrosodimethylamine (NDMA) is a carcinogen found in food, polluted air, amine containing cosmetics and drinking water (*90*). In drinking water NDMA is formed during chlorination process which can be prevented by removal of NDMA precursors prior chlorination or NDMA can be removed with UV treatment (*90*).

In human body NDMA formation is inhibited by various compounds including vitamin C and vitamin E (*91*). Sawamura et al (1999) studied effect of essential oils from 28 different citrus species and individual compounds on inhibition of the formation of NDMA. Essential oil from *Fortunella japonica* Swingle were reported to inhibit 50-70 % of NDMA formation (*92*). Myrcene showed the highest inhibition of 87% of NDMA formation followed by α-terpinene, terpinolene, γ-terpinene, linalool, α-terpineol, α-pinene, β-pinene and citral showing > 50% inhibition of NDMA formation (*92*). Chemical composition of Kumquat oil mainly comprised of limonene (96.7%), myrcene and minor quantities of α-pinene, γ-terpinene, linalool and geranyl acetate (*92*). The inhibitory action of essential oils on NDMA formation is as a result of the synergistic effects of the individual components.

Anti-Microbial Activity

Citrus oils are well known for their anti-microbial activity (*93–95*) and can be utilized as alternatives to the chemical antimicrobials used in food products (*96*). D-Limonene is the main component of kumquat essential oil (*2*, *97*). More recently DL-limonene from *Citrus maxima* and *Citrus sinensis* was reported to possess anti-microbial and antiaflatoxigenic efficacy (*98*). Citral and linalool are considered most effective antimicrobials among the different components of citrus essential oils (*96*, *99*).

In a recent study, kumquat (*Fortunella crassifolia* Swingle) essential oil was demonstrated potential against food pathogenic strains of gram - negative and gram - positive bacteria along with fungus *C. albicans* (*97*). The study tested kumquat essential oil comprising of 25 components on gram-negative bacteria namely *E. coli*, *S. typhimurium* and gram-positive bacteria namely *S.* aureus, *B.* cereus, B. *subtilis*, *L. bulgaricus* and *B. laterosporus*. Gram - positive bacteria had the minimum inhibitory concentration (MIC) values ranging from 37.5 – 67.5 μg/ mL while MIC values for gram - negative ranged between 50 – 70 μg/ mL. The minimum bactericidal concentration (MBC) values for gram - positive bacteria were between 40 – 70 μg/ mL while those for gram - negative bacteria ranged between 53.5 – 72.5 μg/ mL (*97*). Gram - positive bacteria are reported to be more susceptible as compared to gram - negative bacteria to the essential oils (*100*). Minor components such as carveol, carvone, linalool found in Meiwa kumquat were also found to possess antimicrobial activity (*97*). Another use of the kumquat essential oils was reported by Yang et al against four human skin pathogenic micro-organisms (*101*).

In conclusion, this chapter reports the volatile components of kumquats and citrus from various isolation techniques including SDE and HS-SPME. During this process, some artifacts might be formed due to the relative higher temperature. Therefore, HS-SPME method played an important role in determining the quantity of the volatile compounds in food samples. HS-SPME has no solvent peaks to mask the volatile peaks. Kumquats and some citrus species contain alcohols and esters with a p-menthane skeleton. In western citrus species, volatiles constituents reported to have open chain terpenoids such as nerol and geraniol. Further research is needed in identifying the new molecules exhibiting interesting properties. This research will discover minor compounds with potential impact volatiles present in complex and unstudied citrus species using new sophisticated analytical tools. is essential for understanding the knowledge of the flavor and fragrances.

Acknowledgments

These results are based on the work supported by the USDA-NIFA # 2010-34402-20875 "Designing Foods for Health" through the Vegetable & Fruit Improvement Center.

References

1. Reiger, M. Citrus Fruits (Citrus spp.). In *Introduction to Fruit Crops*; Rieger, M., Ed.; Food Products Press: New York, 2006.
2. Koyasako, A.; Bernhard, R. A. *J. Food Sci.* **1983**, *48* (6), 1807–1812.
3. Schirra, M.; Palma, A.; D'Aquino, S.; Angioni, A.; Minello, E. V.; Melis, M.; Cabras, P. *J. Agric. Food Chem.* **2008**, *56* (2), 455–460.
4. Spiegel-Roy, P.; Goldschmidt, E. E. Citrus and Its Relatives. In *Biology of Citrus*; Spiegel-Roy, P., Goldschmidt, E. E., Eds.; Cambridge University Press: Cambridge, U.K., 1996; pp 19–44.
5. Swingle, W. T. The Botany of *Citrus* and Its Wild Relatives of the Orange Subfamily (Family Rutaceae, Subfamily Aurantiodeae). In *Citrus Industry*; Webber, H. J., Batchelor, L., Eds.; University of California: Riverside, CA, 1967; Vol. 1, pp 129–474.
6. Delort, E.; Naef, R. Volatile Components in Less Common Citrus Species. In *Citrus Oils: Composition, Advanced Analytical Techniques, Contaminants, and Biological Activity*; Dugo, G., Mondello, L., Eds.; CRC Press: Boca Raton, FL, 2011; pp 163–191.
7. Ortiz, J. M. Botany: Taxonomy, Morphology and Physiology of Fruits, Leaves and Flowers. In *Citrus: The Genus Citrus*; Dugo, G. G., Di GiacomoA., Eds.; Taylor and Francis: London, 2002; p 16–35.
8. Ye, Y. *Fruit Var. J.* **1985**, *39*, 17–20.
9. Choi, H. S. *J. Agric. Food Chem.* **2005**, *53* (5), 1642–1647.
10. Ladaniya, M. S. Commercial Fresh Citrus Cultivars and Producing Countries. In *Citrus Fruit*; Academic Press, San Diego, 2008; pp 13–V.
11. Park, H. M.; Lee, J. H.; Yaoyao, J.; Jun, H. J.; Lee, S. J. *Biochem. Biophys. Res. Commun.* **2011**, *404* (1), 345–348.
12. Turner, G. W.; Alison, M. B.; Gifford, E. M. *Int. J. Plant Sc.* **1998**, *159* (1), 75–88.
13. Thomson, W. W.; Platt-Aloia, K. A.; Anton, G. E. *Bot. Gaz.* **1976**, *137* (4), 330–340.
14. Bennici, A.; Tani, C. *Flora - Morphol., Distrib., Funct. Ecol. Plants* **2004**, *199* (6), 464–475.
15. Buchanan, B. B. *Biochem. Mol. Biol. Plants* **2000**.
16. Nagegowda, D. A. *FEBS Lett.* **2010**, *584* (14), 2965–2973.

17. Krings, U. *Appl. Microbiol. Biotechnol.* **1998**, *49* (1), 1.
18. Defilippi, B. G.; Manríquez, D.; Luengwilai, K.; González-Agüero, M. Aroma Volatiles: Biosynthesis and Mechanisms of Modulation During Fruit Ripening. In *Advances in Botanical Research*; Kader, J.-C., Delseny, M., Eds.; Academic Press: San Diego, 2009; Vol. 50, Chapter 1, p 1–37.
19. Lucker, J. *Eur. J. Biochem.* **2009**, *269* (13), 3160–3171.
20. Sharon-Asa, L.; Shalit, M.; Frydman, A.; Bar, E.; Holland, D.; Or, E.; Lavi, U.; Lewinsohn, E.; Eyal, Y. *Plant J.* **2003**, *36* (5), 664–674.
21. Miyazawa, M.; Nakamura, Y.; Ishikawa, Y. *J. Agric. Food Chem.* **2000**, *48* (8), 3639–3641.
22. Umano, K.; Hagi, Y.; Tamura, T.; Shoji, A.; Shibamoto, T. *J. Agric. Food Chem.* **1994**, *42* (9), 1888–1890.
23. Hashinaga, F.; Itoo, S. *Bull. Fac. Agric., Kagoshima Univ.* **1990**.
24. Quijano, C. E.; Pino, J. A. *J. Essent. Oil Res.* **2009**, *21*, 194–196.
25. Choi, H.-S. *J. Agric. Food Chem.* **2005**, *53* (5), 1642–1647.
26. Jayaprakasha, G. K.; Murthy, K. N. C.; Demarais, R.; Patil, B. S. *Planta Med.* **2012**, *78* (10), 974–980.
27. Dugo, G. *Citrus Oils: Advanced Analytical Techniques, Composition, and Biological Activity*; CRC Press: Boca Raton, FL 2010; Vol. 49.
28. Bicchi, C. *Flavour Fragrance J.* **2012**, *27* (3), 260–261.
29. van Straten, S.; de Beauveser, J. C. a.; Visscher, C. A. *TNO-CIVO Food Analysis Institute, Zeist, The Netherlands*, 1982.
30. Perez-Cacho, P. R.; Rouseff, R. L. *Crit. Rev. Food Sci. Nutr.* **2008**, *48*, 681–695.
31. Grosch, W. *Chem. Senses* **2001**, *26* (5), 533–545.
32. Bazemore, R.; Goodner, K.; Rouseff, R. *J. Food Sc.* **1999**, *64* (5), 800–803.
33. Hinterholzer, A.; Schieberle, P. *Flavour Fragrance J.* **1998**, *13* (), 49–55.
34. Marin, A. B.; Acree, T. E.; Hotchkiss, J. H.; Nagy, S. *J. Agric. Food Chem.* **1992**, *40* (4), 650–654.
35. Shaw, P. E. *J. Agric. Food Chem.* **1979**, *27* (2), 246–257.
36. Perez-Cacho, P. R.; Rouseff, R. L. *Crit. Rev. Food Sci. Nutr.* **2008**, *48* (7), 681–695.
37. Shaw, P. E. Marcel Dekker, Inc.: New York, 1991.
38. Rega, B.; Fournier, N.; Guichard, E. *J. Agric. Food Chem.* **2003**, *51* (24), 7092–7099.
39. Arena, E.; Guarrera, N.; Campisi, S.; Asmundo, C. N. *Food Chem.* **2006**, *98* (1), 59–62.
40. Buettner, A.; Schieberle, P. *J. Agric. Food Chem.* **1999**, *47* (12), 5189–5193.
41. Flamini, G.; Cioni, P. L. *Food Chem.* **2010**, *120*, 984.
42. Brat, P.; Rega, B.; Alter, P.; Reynes, M.; Brillouet, J.-M. *J. Agric. Food Chem.* **2003**, *51* (11), 3442–3447.
43. Amoore, J. E.; Buttery, R. G. *Chem. Senses* **1978**, *3* (1), 57–71.
44. Nagy, S. a.; Shaw, P. E. *Dev. Food Sci.* **1990**, *3C*, 93–124.
45. Baxter, R. L.; Laurie, W. A.; McHale, D. *Tetrahedron* **1978**, *34* (14), 2195–2199.
46. Marcy, J. E.; Hansen, A. P.; Graumlich, T. R. *J. Food Sci.* **1989**, *54* (1), 227–228.

47. Moshonas, M. G.; Shaw, P. E. *J. Food Sci.* **1989**, *54* (1), 82–85.
48. Caven-Quantrill, D. J.; Buglass, A. J. *J. Chromatogr., A* **2006**, *1117* (2), 121–131.
49. Arthur, C. L.; Pawliszyn, J. *Anal. Chem.* **1990**, *62* (19), 2145–2148.
50. Zhang, Z.; Yang, M. J.; Pawliszyn, J. *Anal. Chem.* **1994**, *66* (17), 844A–853A.
51. Pillonel, L.; Bosset, J. O.; Tabacchi, R. *LWT--Food Sci. Technol.* **2002**, *35* (1), 1–14.
52. Wilkes, J. G.; Conte, E. D.; Kim, Y.; Holcomb, M.; Sutherland, J. B.; Miller, D. W. *J. Chromatogr., A* **2000**, *880* (1–2), 3–33.
53. Čajka, T.; Hajšlová, J.; Cochran, J.; Holadová, K.; Klimánková, E. *J. Sep. Sci.* **2007**, *30* (4), 534–546.
54. Rega, B.; Fournier, N.; Guichard, E. *J. Agric. Food Chem.* **2003**, *51* (24), 7092–7099.
55. Basheer, C.; Wong, W.; Makahleh, A.; Tameem, A. A.; Salhin, A.; Saad, B.; Lee, H. K. *J. Chromatogr., A* **2011**, *1218* (28), 4332–4399.
56. Steffen, A.; Pawliszyn, J. *J. Agric. Food Chem.* **1996**, *44* (8), 2187–2193.
57. Jia, M.; Zhang, Q. H.; Min, D. B. *J. Agric. Food Chem.* **1998**, *46* (7), 2744–2747.
58. Lammers, M.; Dietze, K.; Ternes, W. *J. Food Process. Preserv.* **2011**, *35* (6), 850–860.
59. Flamini, G.; Cioni, P. L.; Morelli, I. *J. Chromatogr., A* **2003**, *998* (1–2), 229–233.
60. Flamini, G.; Tebano, M.; Cioni, P. L. *Anal. Chim. Acta* **2007**, *589* (1), 120–124.
61. Kataoka, H.; Lord, H. L.; Pawliszyn, J. *J. Chromatogr., A* **2000**, *880* (1–2), 35–62.
62. David, F.; Sandra, P. *J. Chromatogr., A* **2007**, *1152* (1–2), 54–69.
63. Charve, J.; Chen, C.; Hegeman, A. D.; Reineccius, G. A. *Flavour Fragrance J.* **2011**, *26* (6), 429–440.
64. Bicchi, C.; D'Amato, A.; Rubiolo, P. *J. Chromatogr., A* **1999**, *843* (1), 99–121.
65. Jayaprakasha, G. K.; Rao, L. J. M.; Sakariah, K. K. *J. Agric. Food Chem.* **2000**, *48* (9), 4294–4295.
66. Jayaprakasha, G. K.; Jena, B. S.; Negi, P. S.; Sakariah, K. K. *Z. Naturforsch., C: J. Biosci.* **2002**, *57* (9–10), 828–835.
67. Rocha, S. M.; Coelho, E.; Zrostlíková, J.; Delgadillo, I.; Coimbra, M. A. *J. Chromatogr., A* **2007**, *1161* (1–2), 292–299.
68. Shellie, R.; Mondello, L.; Marriott, P.; Dugo, G. *J. Chromatogr., A* **2002**, *970* (1–2), 225–234.
69. d'Acampora Zellner, B.; Casilli, A.; Dugo, P.; Dugo, G.; Mondello, L. *J. Chromatogr., A* **2007**, *1141* (2), 279–286.
70. Qiao, Y.; Xie, B. J.; Zhang, Y.; Fan, G.; Yao, X. L.; Pan, S. Y. *Molecules* **2008**, *13* (6), 1333–1344.
71. Bacsik, Z. *Appl. Spectrosc. Rev.* **2004**, *39* (3).

72. Jennings, W.; Shibamoto, T. *Qualitative Analysis of Flavor and Fragrance Volatiles by Glass Capillary Gas Chromatography*; Academic Press: New York, 1980.
73. Dugo, G.; Mondello, L. *Citrus Oils: Composition, Advanced Analytical Techniques, Contaminants, and Biological Activity*; CRC Press, Boca Raton, FL, 2010; Vol. 49.
74. Jabalpurwala, F. A.; Smoot, J. M.; Rouseff, R. L. *Phytochemistry* **2009**, *70* (11–12), 1428–1434.
75. Tu, N. T. M.; Thanh, L. X.; Une, A.; Ukeda, H.; Sawamura, M. *Flavour Fragrance J.* **2002**, *17* (3), 169–174.
76. Tirado, C. B.; Stashenko, E. E.; Combariza, M. Y.; Martinez, J. R. *J. Chromatogr., A* **1995**, *697* (1–2), 501–513.
77. Espina, L.; Somolinos, M.; Lorán, S.; Conchello, P.; García, D.; Pagán, R. *Food Control* **2011**, *22* (6), 896–902.
78. Lota, M.-L.; de Rocca Serra, D.; Tomi, F.; Casanova, J. *Biochem. Syst. Ecol.* **2000**, *28* (1), 61–78.
79. Dugo, G. *J. Essent. Oil Res.* **2011**, *23* (2), 45.
80. Schirra, M.; Agabbio, M.; Continella, G.; D'Aquino, S. *Adv. Hortic. Sci.* **1995**, *6* (2), 83–86.
81. Rodov, V.; Ben-Yehoshua, S.; Albagli, R.; Fang, D. Q. *Postharvest Biol. Technol.* **1995**, *5* (1–2), 119–127.
82. Seifried, H. E. *J. Nutr. Biochem.* **2007**, *18* (3), 168–171.
83. Liu, R. H. *Am. J. Clin. Nutr.* **2003**, *78*, 517S.
84. Choi, H. S.; Song, H. S.; Ukeda, H.; Sawamura, M. *J. Agric. Food Chem.* **2000**, *48* (9), 4156–4161.
85. Murthy, K. N. C.; Jayaprakasha, G. K.; Patil, B. S. *Life Sci.* **2012**, *91*, 429.
86. Patil, J. R.; Jayaprakasha, G. K.; Murthy, K. N. C.; Tichy, S. E.; Chetti, M. B.; Patil, B. S. *Food Chem.* **2009**, *114* (6), 1351–1358.
87. Quijano, C. E.; Pino, J. A. *J. Essent. Oil Res.* **2009**, *21* (6), 483–485.
88. Jayaprakasha, G. K.; M., K. N. C.; Etlinger, M.; Mantur, S. M.; Patil, B. S. *Food Chem.* **2012**, *131* (1), 184–191.
89. Sadek, E.; Makris, D.; Kefalas, P. *Plant Foods Hum. Nutr.* **2009**, *64*, 297.
90. Mitch, W. A.; Sharp, J. O.; Trussell, R. R.; Valentine, R. L.; Alvarez-Cohen, L.; Sedlak, D. L. *Environ. Eng. Sci.* **2003**, *20* (5), 389–404.
91. Ikeda, K.; Migliorese, K. G. *J. Soc. Cosmet. Chem.* **1990**, *41* (5), 283–333.
92. Sawamura, M.; Sun, S. H.; Ozaki, K.; Ishikawa, J.; Ukeda, H. *J. Agric. Food Chem.* **1999**, *47* (12), 4868–4872.
93. Caccioni, D. R. L.; Guizzardi, M.; Biondi, D. M.; Agatino, R.; Ruberto, G. *Int. J. Food Microbiol.* **1998**, *43* (1–2), 73–79.
94. Belletti, N.; Ndagijimana, M.; Sisto, C.; Guerzoni, M. E.; Lanciotti, R.; Gardini, F. *J. Agric. Food Chem.* **2004**, *52* (23), 6932–6938.
95. Viuda-Martos, M.; Ruiz-Navajas, Y.; Fernández-López, J.; Pérez-Álvarez, J. *Food Control* **2008**, *19* (12), 1130–1138.
96. Fisher, K.; Phillips, C. *Trends Food Sci. Technol.* **2008**, *19* (3), 156–164.
97. Wang, Y. W.; Zeng, W. C.; Xu, P. Y.; Lan, Y. J.; Zhu, R. X.; Zhong, K.; Huang, Y. N.; Gao, H. *Int. J. Mol. Sci.* **2012**, *13*, 3382–3393.

98. Singh, P.; Shukla, R.; Prakash, B.; Kumar, A.; Singh, S.; Mishra, P. K.; Dubey, N. K. *Food Chem. Toxic.* **2010**, *48* (6), 1734–1740.
99. Kim, J.; Marshall, M. R.; Wei, C. *J. Agric. Food Chem.* **1995**, *43* (11), 2839–2845.
100. Dabbah, R.; Edwards, V. M.; Moats, W. A. *Appl. Microbiol.* **1970**, *19* (1), 27–31.
101. Yang, E. J.; Kim, S. S.; Moon, J. Y.; Oh, T. H.; Baik, J. S.; Lee, N. H.; Hyun, C. G. *Acta Microbiol. Immunol. Hung.* **2010**, *57* (1), 15–27.

Health-Promoting Properties of Fruits

Chapter 5

Antioxidant Activity of Anthocyanin-Rich Colombian Tropical Fruits

Coralia Osorio*,[1] and Ovidio Almanza[2]

[1]Departamento de Química and Universidad Nacional de Colombia, AA 14490 Bogotá, Colombia
[2]Departamento de Física, Universidad Nacional de Colombia, AA 14490 Bogotá, Colombia
***E-mail: cosorior@unal.edu.co.**

Total anthocyanin content of five Colombian tropical fruits was measured for the first time, and their antioxidant activity was determined by using EPR (Electron Paramagnetic Spectroscopy). A second-order kinetic model was applied to the free radical scavenging reaction under presence of ABTS and DPPH free radicals to compare their antioxidative potential. The total monomeric anthocyanin content of corozo (*Bactris guineensis*), motilón (*Hyeronima macrocarpa* Müll. Arg.), coral (*Hyeronima macrocarpa* Müll. Arg.), "uva de árbol" (*Myrciaria* aff. *cauliflora*), and "mora pequeña" (*Rubus megalococcus* Focke.) was determined by the pH-differential method. The anthocyanin-rich extracts of motilón and corozo exhibited not only the highest anthocyanin content but also the highest antioxidant potential.

Antioxidants are compounds that restrain or delay the oxidation of other molecules by inhibiting the initiation or propagation of oxidizing chain reactions. Among them, anthocyanins are considered to contribute to the beneficial effects of consuming fruits and vegetables (*1*), and various anthocyanin-containing extracts from fruits have been shown to reduce the oxidative stress-associated

© 2013 American Chemical Society

to inflammatory diseases and cancer (*2–5*), and also to be a potential candidates as a complementary photoprotective agent of skin (*6*). Thus, the association between a diet rich in fruit and vegetables and a decreased risk of cardiovascular diseases, cancer, and immune system decline has been supported by considerable epidemiological evidence (*7*, *8*).

Electron paramagnetic resonance (EPR) is a valuable tool for detecting and identifying free radicals generated by *ex vivo* or *in vivo* chemical reactions. The basis of this spectroscopical technique is the absorption of microwave energy by unpaired electrons, when they are in a magnetic field. The free radical scavenger activity of phytochemicals is measured by their ability to react with free radicals present in the same solution. There are two different antioxidative parameters to be determined during the above-mentioned reaction, the antioxidative potential, that represents the velocity of the radical degradation (kinetic phenomenon, μmol degraded radical per minute), and the antioxidative capacity that is the amount of radical which is degrade after a certain time by a given concentration of antioxidant (μmol degraded radical per μmol antioxidant) (*9*). Both of them are easily measured by EPR, a highly specific and sensitive technique, such obtaining valuable information over the antioxidant capacity of phytochemicals under evaluation.

Colombia has a wide variety of exotic fruits. Up to now, there is only some available data concerning the main constituents (sugar, acids) and also about volatile composition. So, knowledge regarding to antioxidant activity of Colombian fruits will provide important information needed to open new markets for these products. As part of our ongoing research, the anthocyanin-rich extracts (AREs) of five Colombian fruits were obtained and their composition determined by spectroscopical techniques (LCMS and NMR). The selected fruits were: corozo (*Bactris guineensis*) (*10*), "mora pequeña" (*Rubus megalococcus* Focke), "uva de árbol" (*Myrciaria* aff. *cauliflora* (Mart) O. Berg), coral (*Hyeronima macrocarpa* Müll. Arg.), and motilón (*Hyeronima macrocarpa* Müll. Arg.) (*11*).

Thus, the aim of this work was to to determine the antioxidant activity of each ARE under the presence of ABTS and DPPH free radicals by EPR, taking into account their high-content of anthocyanins and other phenolic compounds.

Experimental Procedures

Fruits

Fruits of "mora pequeña" (*R. megalococcus* Focke) were harvested in Albania (Santander, Colombia), "uva de arbol" (*M.* aff. *cauliflora* (Mart.) O. Berg) and coral (*H. macrocarpa* Müll. Arg.) in Timbio (Cauca, Colombia), motilón (*H. macrocarpa* Müll. Arg.) in La Unión (Nariño, Colombia), and corozo (*B. guineensis*) in Montería (Córdoba, Colombia). All fruits were randomly selected and the maturity stage checked by pH and °Brix measurements. Upon arrival at the Universidad Nacional de Colombia, the samples were frozen and then stored at -23°C until further analyses.

Chemicals

2,2′-Azino-bis(3-ethylbenzothiazoline-6-sulfonic acid) (ABTS), and 1,1-diphenyl-2-picrylhydrazyl (DPPH), were purchased from Sigma Chemical Co. (Steinheim, Germany). 6-Hydroxy-2,5,7,8-tetramethylchroman-2-carboxylic acid (Trolox) was obtained from Fluka Chemie GmbM (Steinheim, Germany). Potassium persulfate was obtained from Merck (Darmstadt, Germany).

Anthocyanin Content

Fruit samples were powdered in liquid nitrogen using a blender, and the extracts were obtained following the method described by Rodríguez-Saona *et al.* (*12*). Briefly, a sample (*ca.* 50 g) of each fruit was powdered in liquid nitrogen and then extracted with acetone. Then the extracted material was separated from the cake by vacuum filtration. The filter cake was extracted again with aqueous acetone (70%, *v/v*). Filtrates were combined and partitioned with chloroform (1:2, acetone: chloroform, *v/v*). The aqueous fraction was collected, concentrated under vacuum and redisolved in acidified water until a volume of 250 mL. Sample extractions were replicated three times.

Total anthocyanin content (TAC) was determined by using the pH differential method described by Giusti and Wrolstad (*13*, *14*), based on total anthocyanin transformation to flavylium cation after a pH decreasing. TAC was expressed as cyanidin-3-glucoside using an extinction coefficient (ε) of 26900 L/mol.cm and molecular weight of 449.2.

Antioxidant Activity

For this purpose, 0.5 g of each freeze-dried fruit were extracted with 20 mL of methanol: water (1:1, *v/v*) and subsequently with 20 mL of acetone:water (7:3, *v/v*). The supernatants were combined in 50 mL volumetric flasks (*15*). Extractions were performed in triplicate. The radical scavenging activity of each fruit extract against ABTS and DPPH free radicals was measured by EPR. For ABTS assay, the free radical solution was prepared according to the method published by Re *et al.* (*16*).

EPR measurements were performed in a Bruker ESP 300 spectrometer at room temperature, by using a radiation of 9.44 GHz (X band) with a modulation frequency of 100 kHz, a sweep width of 100 G, modulation amplitude of 0.49 G, scan time of 41.94 s, and microwave power of 20 mW. The integral intensities of EPR spectra were obtained by evaluating their double integrals (DI_{EPR}). The relative concentration (C_{rel}) of free radicals for different reaction times were calculated by relation to the double integral of reference (free radical solution) (*17*). A second-order kinetic model was assumed for the free radical scavenging reaction of AREs against ABTS and DPPH free radicals (*18*). Antioxidant potential was determined by plotting the expression $1/C_t - 1/C_0$ against time for each fruit, and further linear regression analysis that allow to calculate the slope (rate constant, *K*).

Results and Discussion

Five native and wild red fruits were selected for this study. Their anthocyanin composition was already analyzed by spectroscopical methods (NMR of pure compounds and LCMS analyses in comparison with standards). A summary of the anthocyanin composition of fruit extracts, based on our previously reported results, is presented in Table 1. Motilón (*Hyeronima macrocarpa* Müll. Arg.) and coral (*Hyeronima macrocarpa* Müll. Arg.) exhibited a similar qualitative and quantitative composition, with delphinidin-3-*O*-rutinoside being the major constituent (*11*). "Uva de árbol" (*Myrciaria* aff. *Cauliflora* (Mart) O. Berg) and "mora pequeña" have in common a high content of cyanidin-3-*O*-glucoside (higher than 80% of anthocyanin-rich extract) (*11*) and the other fruit, corozo (*Bactris guineensis*) was composed mainly by cyanidin-3-*O*-rutinoside and cyanidin-3-*O*-glucoside (*ca.* 90% of anthocyanin-rich extract) (*10*).

The quantification of anthocyanins in each fruit extract was made pH-differential method. This is an inexpensive and effective method to quantify the total anthocyanin content in a sample, which has been approval by AOAC (Association of Analytical Communities) and presented a good correlation with quantification of anthocyanins by HPLC (*14*). As it can be seen in Table 1, motilón and corozo have the highest anthocyanin concentration. This result is remarkable, taking into account that red color is only present in the skin of these fruits. In contrast, "mora pequeña" that is a red-fleshed berry showed a lower anthocyanin-content in comparison with the above-mentioned fruits.

The measurement of antioxidant activity of AREs was done by EPR. This is an useful technique because the intensity of the EPR spectra is proportional to the number of free radicals in solution. Thus, an intensity decrease is expected in the presence of antioxidant compound(s) that could scavenge the initial free radicals or inhibit the formation of secondary forms. The most remarkable advantage of this technique in comparison to the usual UV-Vis spectroscopy is the direct measurement of free radical amount.

The reaction kinetics for each ARE, in the presence of ABTS (Figure 1) or DPPH (Figure 2) free radicals, were plotted. The stability of free radical solutions were confirmed by monitoring their concentration through the time (reference). The evolution of the different reaction kinetics depends on the nature of antioxidant compounds present in each fruit extract, thus, differences among fruits extracts and also free radicals were obtained. In both of the cases, the most active extracts were those of motilón and corozo. It is important to point out that corozo ARE abated entirely the DPPH free radicals in solution. These results were expressed as antioxidative potential, by application of a second-order kinetic model and calculation of rate constant (*K*) in each case. From these results, the lowest activity of "mora pequeña" anthocyanin-rich extract was evident. In all of the cases (with exception for coral and "mora pequeña" against DPPH), the concentration of free radicals decrease less than 50% in 3 minutes. This behavior is desirable to avoid the damages produced by the reactive free radicals.

Table 1. Anthocyanin Content and Antioxidant Potential (Free-Radical Scavenging Activity Rate Constants) of Fruit Extracts Against ABTS and DPPH

Fruit	*Total monomeric anthocyanin content (mg/100 g FW)*[a]	*Main anthocyanins (%)* [b]								*Antioxidant potential K (L/mol.min)*	
		C3G	*C3R*	*C3M*	*C3S*	*D3G*	*D3R*	*P3R*	*PN3R*	*ABTS*	*DPPH*
Motilón (*Hyeronima macrocarpa*)	239.92 ± 3.43	-	25.6	-	-	9.7	37.0	27.7	-	1.54×10^7	1.17×10^7
Corozo (*Bactris guineensis*)	80.02 ± 1.05	15.7	72.2	4.1	1.8	-	-	-	5.1	2.02×10^6	1.61×10^8
Uva de árbol (*Myrciaria* aff. *cauliflora*)	55.61 ± 3.39	84.5	-	-	-	15.5	-	-	-	1.93×10^6	1.36×10^6
Mora pequeña (*Rubus megalococcus*)	25.52 ± 0.10	84.1	1.1	14.8	-	-	-	-	-	7.71×10^5	5.85×10^5
Coral (*Hyeronima macrocarpa*)	22.74 ± 1.38	-	12.0	-	-	3.4	70.1	14.5	-	1.73×10^6	5.85×10^5

[a] Expressed as cyanidin-3-glucoside. [b] Data taken from literature (*10, 11*). C3G = cyanidin-3-*O*-glucoside, C3R = cyanidin-3-*O*-rutinoside, C3M = cyanidin-3-*O*-(6″-malonyl)-glucoside, C3S = cyanidin-3-sambubioside, D3G = delphinidin-3-*O*-glucoside, D3R = delphinidin-3-*O*-rutinoside, P3R = petunidin-3-*O*-rutinoside, PN3R = peonidin-3-*O*-rutinoside.

The higher free-radical scavenging activity could be related to the amount of total monomeric anthocyanin content, because motilón and corozo exhibited not also the highest antioxidant potential but also the highest TAC, among studied fruits. It seems that the qualitative anthocyanin composition of motilón and coral AREs do not have any influence on the antioxidant potential, taking into account that the same anthocyanins were found in them (*11*). For the case of corozo, it was probed the synergistic or cooperative effects of anthocyanin mixtures as responsible of the highest antioxidant activity of this ARE in comparison to pure compounds (*17*).

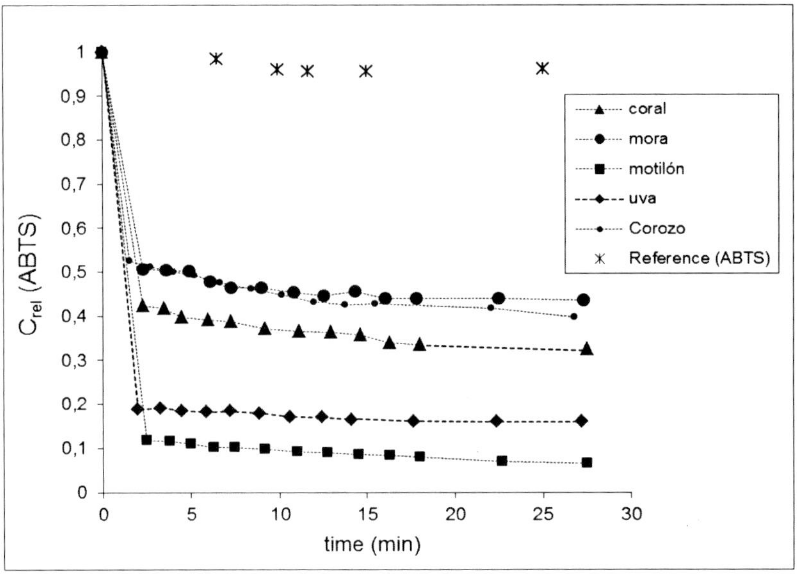

Figure 1. Relative concentration of ABTS free radical under the presence of anthocyanin-rich extracts of fruits.

The anthocyanin-rich fruit extracts usually has been considered as source of antioxidants that scavenge free radicals and therefore, important tools for preserving good health (*19*, *20*). However, recent evidence suggests that ROS (reactive oxygen species) exert essential metabolic functions and their entirely removal can upset cell signaling pathways and opposite to the expectations, increase the risk of chronic disease. This has been to take into account by food industry during the use of antioxidants (*21*).

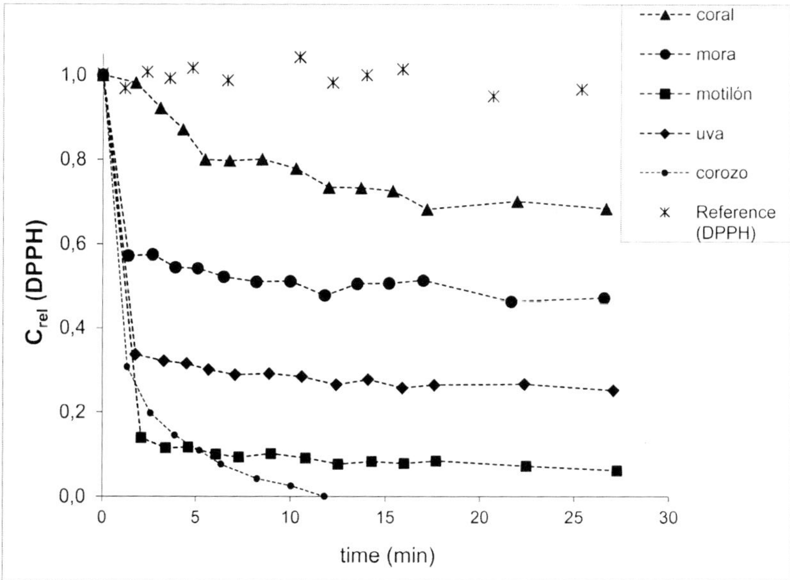

Figure 2. Relative concentration of DPPH free radical under the presence of anthocyanin-rich extracts of fruits.

Conclusion

This work illustrates the application of EPR in the study of antioxidant activity of anthocyanin-rich extracts of five wild Colombian tropical fruits. These results allow to choose two promising exotic fruits as source of natural colorants exhibiting antioxidant properties. It is expected that these fruit extracts will be able to use in the food industry after their characterization and the achievement of a stability storage test.

Acknowledgments

The authors gratefully acknowledge the financial support from DIB-Universidad Nacional de Colombia and IFS (International Foundation for Science), Sweden (project E/4826-1).

References

1. Wang, H.; Cao, G.; Prior, R. L. *J. Agric. Food Chem.* **1997**, *45*, 304–309.
2. Kong, J.-M.; Chia, L.-S.; Goh, N.-K.; Chia, T.-F.; Brouillard, R. *Phytochemistry* **2003**, *64*, 923–933.
3. Chen, P.-N.; Chu, S.-Ch.; Chiou, H.-L.; Kuo, W.-H.; Chiang, Ch.-L.; Hsieh, Y.-S. *Cancer Lett.* **2006**, *235*, 248–259.
4. Dai, J.; Gupte, A.; Gates, L.; Mumper, R. J. *Food Chem. Toxicol.* **2009**, *47*, 837–847.

5. Barrios, J.; Cordero, C. P.; Aristizabal, F.; Heredia-Mira, F. J.; Morales, A. L.; Osorio, C. *J. Agric. Food Chem.* **2010**, *58*, 2100–2110.
6. Cimino, F.; Ambra, R.; Canali, R.; Saija, A.; Virgili, F. *J. Agric. Food Chem.* **2006**, *54*, 4041–4047.
7. Michels, K. B.; Giovannucci, E.; Joshipura, K. J.; Rosner, B. A.; Stampfer, M. J.; Fuchs, C. S.; Colditz, G. A.; Speizer, F. E.; Willet, W. C. *J. Natl. Cancer Inst.* **2000**, *92*, 1740–1752.
8. Aiyer, R. S.; Warri, A. M.; Woode, D. R.; Hilakivi-Clarke, L.; Clareke, R. *J. Agric. Food Chem.* **2012**, *60*, 5693–5708.
9. Rohn, S.; Kroh, L. W. *Mol. Nutr. Food Res.* **2005**, *49*, 898–907.
10. Osorio, C.; Acevedo, B.; Hillebrand, S.; Carriazo, J.; Winterhalter, P.; Morales, A. L. *J. Agric. Food Chem.* **2010**, *58*, 6977–6985.
11. Santacruz, L.; Carriazo, J. G.; Almanza, O.; Osorio, C. *J. Agric. Food Chem.* **2012**, *60*, 1397–1404.
12. Rodríguez-Saona, L. E.; Wrolstad, R. E. In *Current Protocols in Food Analytical Chemistry*; Wrolstad, R. E., Acree, T. E., An, H., Decker, E. A., Penner, M. H., Reid, D. S., Schwartz, S. J., Shoemaker, C. F., Sporns, P., Eds.; John Wiley & Sons, Inc.: New York, 2001; pp F1.1.1−F1.1.11.
13. Giusti, M. M.; Wrolstad, R. E. In *Current Protocols in Food Analytical Chemistry*; Wrolstad, R. E., Acree, T. E., An, H., Decker, E. A., Penner, M. H., Reid, D. S., Schwartz, S. J., Shoemaker, C. F., Sporns, P., Eds.; John Wiley & Sons, Inc.: New York, 2001; pp F1.2.1−F1.2.13.
14. Lee, J.; Rennaker, C.; Wrolstad, R. E. *Food Chem.* **2008**, *110*, 782–786.
15. Vasco, C.; Ruales, J.; Kamal-Eldin, A. *Food Chem.* **2008**, *111*, 816–823.
16. Re, R.; Pellegrini, N.; Proteggente, A.; Pannala, A.; Yang, M.; Rice-Evans, C. *Free Radical Biol. Med.* **1999**, *26*, 1231–1237.
17. Osorio, C.; Carriazo, J. G.; Almanza, O. *Eur. Food Res. Technol.* **2011**, *233*, 103–108.
18. Rufino, M. S. M.; Alves, R. E.; Fernandes, F. A. N.; Brito, E. S. *Food Res. Int.* **2011**, *44*, 2072–2075.
19. Kong, J.-M.; Chia, L.-S.; Goh, N.-K.; Chia, T.-F. *Phytochemistry* **2003**, *64*, 923–933.
20. Jensen, G. S.; Wu, X.; Patterson, K. M.; Barnes, J.; Carter, S. G.; Scherwitz, L.; Beaman, R.; Endres, J. R.; Schauss, A. G. *J. Agric. Food Chem.* **2008**, *56*, 8326–8333.
21. Finley, J. W.; Kong, A-N.; Hintze, K. J.; Jeffery, E. H.; Ji, L.; Lei, X. G. *J. Agric. Food Chem.* **2011**, *59*, 6837–6846.

Chapter 6

Flavonoid and Antioxidant Properties of Fruits Belonging to the *Annona* and *Citrus* Genera

Davide Barreca,* Ersilia Bellocco, Corrado Caristi, Ugo Leuzzi, and Giuseppe Gattuso

Dipartimento di Chimica Organica e Biologica, Università di Messina, Viale F. Stagno d'Alcontres 31, 98166 Messina, Italy
***E-mail: dbarreca@unime.it. Telephone: +39 090 6765187. Fax: +39 090 6765186.**

The present chapter describes the investigations on the composition, antioxidant activity and biological properties of two tropical/subtropical fruits, namely kumquat (*Citrus japonica*) and cherimoya (*Annona cherimola*). Kumquat juice and cherimoya pulp extract showed remarkable antioxidant power and biological activity.

Introduction

Over the past few years, pharmaceutical research has progressively shifted its focus from fully synthetic compounds to natural products, implicitly recognizing that there is still a huge pool of bioactive compounds whose properties (and chemical structures, in many cases) have not yet been revealed. This is the consequence of the growing awareness that the plant kingdom is possibly the main source of health-beneficial compounds, whose intake mostly come from a diet rich in fruit and vegetables.

As a matter of fact, it has been widely accepted by the scientific community that diet and health are deeply intertwined (*1*, *2*). Aside from their richness in macronutrients fundamental for sustainment (e.g., carbohydrates, proteins or

© 2013 American Chemical Society

fats), fruits and vegetables possess a variety of different micronutrients, ranging from inorganic (minerals) to organic ones such as vitamins and phenolics. Many of these compounds play key roles in metabolic processes providing benefits that go beyond the mere correct biochemical functioning of the human body. *In vitro*, *in vivo* and epidemiological studies have shown that specific micronutrients may also provide additional protective effects against the insurgence of several diseases. As a result, the World Health Organization (WHO) recently stated that the intake of 400g/day of fruits and vegetables may provide a fair level of protection against non-transmissible chronic pathologies such as cardiovascular and neurodegenerative diseases or some types of cancer.

These pathologies are often related to the action of 'free' radicals, which are generated *in vivo* as Reactive Oxygen Species (ROS): singlet oxygen, peroxynitrite, superoxide, peroxyl and hydroxyl radical, when present in too high concentration, may put cells under oxidative stress and cause serious damage. Antioxidant compounds influence this delicate balance, adding their activity to the innate defense system of the human body.

Many of the microcomponents responsible for this preventing action belong to the polyphenols group. Among these, in particular, the broad family of flavonoids is receiving considerable attention, given that they are ubiquitous in fruits and vegetables and show remarkable antioxidant activity, that is, they are able to intercept and quench ROS (*3*, *4*).

Flavonoids are plant secondary metabolites formally derived from 2-phenyl-1,4-benzopyrone. They are further classified based on structural variation mainly on the (central) pyrone ring (*5*). The common feature of most flavonoids, however, is the presence of a number of phenolic OH groups. These groups are responsible for the antioxidant activity, as they are able to transfer atomic hydrogen ($H^{\bullet}$) to more reactive free radicals, yielding as a result stable radical species that do not react further (*6*).

An interesting point is that the various flavonoid subclasses occur with a very different distribution in plant species and, owing to the fact that more than 5,000 flavonoid derivatives have been identified so far, they have often been used as chemotaxonomic markers. As a matter of fact, the chromatographic distribution of the various components may well be regarded as a 'fingerprint', to identify the origin of a given fruit-derived product (*7*), or to confirm (or even reveal) taxonomical relations between different plant species.

As mentioned above, fruits and vegetables are the main dietary sources of phenolics. Many of the most common species, that is, those universally distributed around the world, have been studied in detail, and plenty of information is available in the literature. However, there is still a huge variety of species that is grown locally – especially in the tropical and sub-tropical regions for reasons related to climate or tradition – that have not been studied in sufficient detail, both on the analytical and biological point of view. About these, even though they are often extremely important for the sustainment of indigenous populations, little is known, both in terms of composition and potential health benefits. The present chapter will be focused on two such fruits, cherimoya (*Annona cherimola* Mill.) and kumquat (*C. japonica* Thunb.).

Results and Discussion

Kumquat (*C. japonica* Thunb.)

For several decades, kumquats have been classified, within the family Rutaceae, into a different genus than *Citrus*, i.e. *Fortunella*. Much of the literature on this species, in fact, considers kumquats as few separate species (*F. japonica*, *F. crassifolia*, *F. hindsii*, *F. japonica*, *F. margarita*, *F. obovata*, *F. polyandra*). Recognizing that it is closely related to *Citrus*, very recently (*8*) kumquat has been reclassified, reinstating the older binomial name *Citrus japonica*, which now encompasses all the *Fortunella* species, considering them as varieties (cultivars) of *C. japonica*.

Kumquat, native to South Eastern Asia (China), is considered to be one of the five ancestral *Citrus* species (*9*) from which all the present species have originated. It is a prolific bushy tree, which grows oval or round-shaped fruits with a smooth, bright orange rind, with a diameter of about 1.0–1.2 cm and 3–3.5 cm long. Differently from most *Citrus*, it is consumed whole (i.e., including the peelings).

The quantitative and qualitative flavonoid composition of kumquat juice obtained from fruits at different stages of ripening (*10*) has been elucidated, focusing the attention also on nonedible unripe fruit – basing the distinction on color, firmness and size – as a potential source of bioactive compounds. Investigations have been carried out by applying an analytical protocol we have optimized over the years for the study of the flavonoid and furocoumarin fractions of juices obtained from *Citrus* species. This protocol allows, in a single chromatographic course, for the qualitative and quantitative determination of the analytes by reverse-phase HPLC with DAD detection, coupled with ESI MS-MS (*11*), by means of a direct injection of freshly squeezed crude juice (diluted with DMF, 1:1 v/v). DAD detection at three different wavelengths – 278, 310 and 325 nm – permits to preliminarily define whether an individual analyte possesses a flavone, a flavanone or a furocoumarin skeleton. In fact, flavones possess two typical UV-vis maxima at 320–335 nm (the so-called band I) and 270–290 nm (band II) assignable, respectively, to the cinnamoyl moiety composed of ring B and the 2,3-double bond of ring C, and to the ring A benzoyl moiety, whereas flavanones show a band II absorption at 270–295 nm, but and a shoulder of lower intensity in place of band I. Furocoumarins, on the other hand, possess an absorption maximum around 310 nm, as well as producing characteristic (and easily distinguishable from those of flavonoids) UV-vis spectra with maxima around 220, 250 and 270 nm.

Flavonoids in *Citrus* occur mostly as glycosyl derivatives, with the sugar moieties linked to the aglycone by means of *O*- or *C*-glycosidic bonds (see below). A parallel treatment of a juice sample with HCl (*12*) comes in handy to obtain further information: *O*-glycosides are cleaved under these conditions, whereas *C*-glycosides are not. Comparison of the HPLC trace of crude and HCl-treated juice provides leading evidence on the type of glycosidic bond present in the individual analytes. ESI-MS-MS, carried out both in positive and negative ion mode, finally allows to identify the nature and the position of the saccharide substituent(s) (*13*).

Quantification of the individual components can be conveniently carried out by Selected Reaction Monitoring (SRM).

Citrus juices are generally characterized by the abundance of flavanone and flavone derivatives, the former being often fewer but quantitatively more abundant, the latter being present in lower amount but in larger variety (*14–16*). As mentioned above, both flavanones and flavones occur mostly as glycosyl derivative, bearing rutinose (α-(1→6)-L-rhamnopyranosyl-β-D-glucopyranose), neohesperidose (α-(1→2)-L-rhamnopyranosyl-β-D-glucopyranose) and/or glucose as *O*-linked substituents. In addition, flavones may be found as mono- or di-*C*-linked glucosyl derivatives. Different species, however, yield different distributions: orange (*17*) and tangerine, for instance, are characterized by the abundance of hesperidin and, to a less extent, narirutin, lemon by hesperidin, eriocitrin and diosmin, Mediterranean sweet lemon (*18*) by hesperidin and eriocitrin, and grapefruit by naringin and narirutin (*19*).

Bergamot (*20*, *21*), sour orange (*22*) and myrtle-leave orange (*23*, *24*), on the other hand, contain the bitter-tasting neohesperidosides, i.e., naringin, neohesperidin and neoeriocitrin, as the predominant compounds.

Kumquat juice, however, by applying the identification/quantification protocol described above, was found to differ sensibly from those obtained from the juices from the other *Citrus* species analyzed (Figure 1) (*10*). In fact, the pre-eminent component was found to be phloretin 3′,5′-di-*C*-glucoside **6**, a dihydrochalcone derivative (the only member of this class of compounds found in kumquat juice) – formally derived by a flavanone skeleton with a 'cleaved' ring C – bearing two glucose units linked via *C*-glycosidic bonds to ring A, which was quantified as 62.49 ± 0.50 and 19.94 ± 0.50 mg/L in the juice from unripe and ripe kumquat, respectively (*10*). Alongside, a further twelve components were found, five of which for the first time. Eight compounds were assigned a flavone skeleton (acacetin 3,6-di-C-glucoside (**1**), vicenin-2 (**2**), lucenin-2 4′-methyl ether (**3**), apigenin 8-*C*-neohesperidoside (**5**), rhoifolin (**8**), acacetin 8-*C*-neohesperidoside (**9**), acacetin 8-C-neohesperidoside (**10**) and acacetin 7-*O*-neohesperidoside (**12**), Table I), whereas the remaining four were found to belong to the flavanone class (narirutin 4′-*O*-glucoside (**4**), hesperidin (**7**), didymin (**11**) and poncirin (**13**), Table II). All these coumpound were present in much lower amount than phloretin **6** (< 2.8 mg/L each in juice from unripe fruits, and > 0.85 mg/L each in juice from ripe ones (*10*)).

In the course of this study, the flavonoid content of juices obtained from fruits at a different ripening stage was determined. It was found that the qualitative flavonoid profile in the juices from green and ripe kumquat did not differ. However, juice from unripe fruits was found to be *ca.* three times higher than that determined for the juice from ripe fruits (71.2±0.09 *vs.* 23.1±0.19 mg/L, respectively (*10*)).

The chromatographic separation, identification and quantification of the juice components were of fundamental importance for the following antioxidant activity assays. With the aim of gaining additional information on the constituents responsible of such activity, along with the antioxidant activity of the whole crude juice, some key juice fractions were assayed. Preparative HPLC (C-18 reverse phase, linear gradient of acetonitrile (5-100%)) led to the separation of three groups of compounds (Figure 2): (i) a broad fraction containing the entire flavonoid pool (FP), (ii) a fraction containing only the main flavonoid component, i.e. the phloretin derivative **6** (X), and (iii) a fraction containing the remaining minor flavonoid components (FP–X). This procedure was applied to juices obtained from both ripe and unripe fruits (prefixes R and U, respectively, in Figure 2). The solutions thus obtained, along with samples of freshly squeezed crude juice, were used for the determination of antioxidant activity, which was assessed by quenching of two stable free radicals, 2,2-diphenyl-1-picrylhydrazyl (DPPH$^{\bullet}$) and 2,2′-azino-bis (3-ethylbenzothiazoline-6-sulphonic acid) radical cations (ABTS$^{\bullet +}$). Results are reported in Figure 3.

Both juices showed a marked antioxidant activity (*10*), bleaching DPPH$^{\bullet}$ radicals, although the juice obtained from ripe fruits showed twice the activity with respect to the one from green fruits (RJ: 14.73 and UJ: 8.57 μM Trolox Equivalents (TE), respectively). Further observations could be gathered from the comparison of the activity assessed for the flavonoid pool FP fractions (R-FP and U-FP). The U-FP fraction (6.83 μM TE) accounts for ~79% of the activity of crude unripe juice (UJ), whereas the R-FP fraction (5.73 μM TE) accounts for only ~34% of the activity of crude ripe juice (RJ) (*10*). In addition, the activity measured for the fraction containing phloretin 3′,5′-di-*C*-glucoside **6** (R-X and U-X fractions) showed that the phloretin derivative in unripe fruit juice is responsible for ~57% (3.91 and 2.29 μM TE) of the activity of the U-FP fraction whereas, in the case of ripe fruits, R-X (2.29 μM TE) accounts for ~40% of the total activity of the R-FP fraction (*10*).

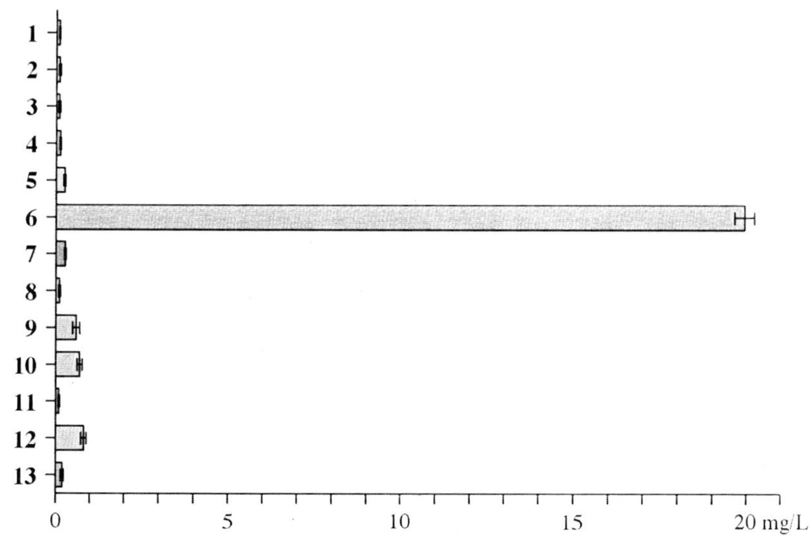

Figure 1. Flavonoids distribution in kumquat (C. japonica) juice.

Table I. Flavone-*C*-Glucosides (1-3, 5, 9 and 10) and Flavone-*O*-Glycosides (8 and 12) (Trivial names are indicated in parentheses)

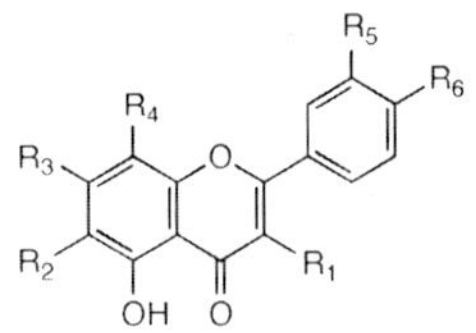

	R_1	R_2	R_3	R_4	R_5	R_6	Structure assignment
1	Glu	Glu	OH	H	H	OMe	Acacetin 3,6-di-*C*-glucoside
2	H	Glu	OH	Glu	H	OH	Apigenin 6,8-di-*C*-glucoside (Vicenin-2)
3	H	Glu	OH	Glu	OH	OMe	Diosmetin 6,8-di-*C*-glucoside (Lucenin-2 4'-methyl ether)
5	H	H	OH	Nh[a]	H	OH	Apigenin 8-*C*-neohesperidoside
8	H	H	*O*-Nh[a]	H	H	OH	Apigenin 7-*O*-neohesperidoside (Rhoifolin)
9	H	H	OH	Nh[a]	H	OMe	Acacetin 8-*C*-neohesperidoside (2"-*O*-rhamnosyl cytisoside)
10	H	Nh[a]	OH	H	H	OMe	Acacetin 6-*C*-neohesperidoside (2"-*O*-rhamnosyl isocytisoside)
12	H	H	*O*-Nh[a]	H	H	OMe	Acacetin 7-*O*-neohesperidoside (Fortunellin)

[a]*O*-Neohesperidose.

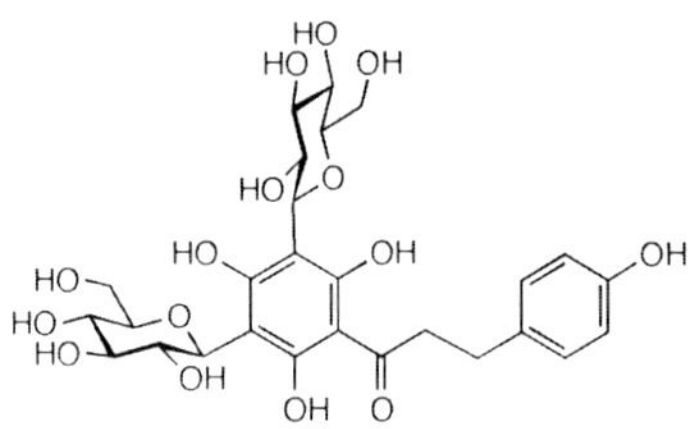

6 Phloretin 3',5'-di-*C*-glucoside

As far as the minor flavonoids fractions (R-FP–X and U-FP–X) it is worth noticing that, even if they are quantitatively less important that the main phloretin component, they show a higher activity (3.22 and 4.63 μM TE for the R-FP–X and the U-FP–X fractions, respectively) in bleaching DPPH• radicals, therefore contributing in a significant manner to the antioxidant power determined by this assay.

Kumquat juice was found to be even more efficient in quenching ABTS•+ radical cations than DPPH• radicals (RJ: 20.78 μM TE; UJ: 19.35 μM TE), although in this case no striking difference was observed between the juice from ripe and unripe kumquats. Once again interesting results derived by the comparison of the antioxidant activity of the crude juices with those measured for the corresponding fractions. For RJ, the total activity is much higher than that shown by the flavonoid pool R-FP fraction (8.0 μM TE), and the latter is expressed mostly by the phloretin component **6** (6.7 μM TE) rather than by the remaining flavonoids (1.8 μM TE). Conversely, in the case of UJ most of the activity derives from the flavonoid pool U-FP fraction (17.98 μM TE), although yet again within the flavonoid pool it is compound **6** (14.54 μM TE) that provide most of the scavenging ability.

Cross comparisons between the different set of experiments provided additional deductions: in general, in unripe fruits the flavonoid pool provide most of the action against both free radicals (~79% *vs.* $DPPH^{\bullet}$ and ~93% *vs.* $ABTS^{\bullet+}$), whereas in the juice of ripe fruits the flavonoid pool contributes for *ca.* one third of the total activity (~34% and ~38% against $DPPH^{\bullet}$ and $ABTS^{\bullet+}$, respectively). This observation strongly suggests that upon ripening, some other component of the juice strongly contributes to the activity (*25*). This is also backed up by the observation that, in line with the different total flavonoid content of the two juices (as mentioned above, unripe fruits contain *ca.* three times as much flavonoids than ripe ones), U-FP fractions display higher activity than R-FP fractions, even though the juice from ripe fruits (RJ) as a whole is more efficient than the juice from unripe fruits (UJ).

A second conclusion can be drawn by looking at the relative contribution of X fractions (i.e., phloretin **6**) and FP–X fractions (i.e., flavones and flavanones): there is an evident trend showing that the FP–X are better $DPPH^{\bullet}$ scavengers than X fractions (in both ripe and unripe juices), whereas this behavior is inverted when $ABTS^{\bullet+}$ scavenging is concerned. This may depend on the different mechanisms involved in $DPPH^{\bullet}$ and $ABTS^{\bullet+}$ inactivation (*6, 26*).

Table II. Flavanone-*O*-Glycosides (4, 7, 11 and 13) (Trivial names are indicated in parentheses)

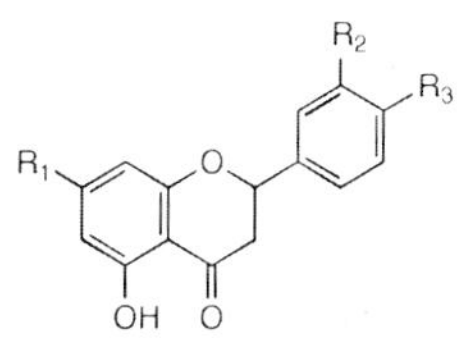

	R_1	R_2	R_3	Structure assignment
4	*O*-Ru[a]	H	*O*-Glu	Naringenin 7-*O*-rutinoside-4′-*O*-glucoside (Narirutin 4′-*O*-glucoside)
7	*O*-Ru[a]	OH	OMe	Hesperetin 7-*O*-rutinoside (Hesperidin)
11	*O*-Ru[a]	H	OH	Isosakuranetin 7-*O*-rutinoside (Didymin, Neoponcirin)
13	*O*-Nh[b]	OH	OMe	Isosakuranetin 7-*O*-neohesperidoside (Poncirin)

[a]*O*-Rutinose; [b]*O*-Neohesperidose

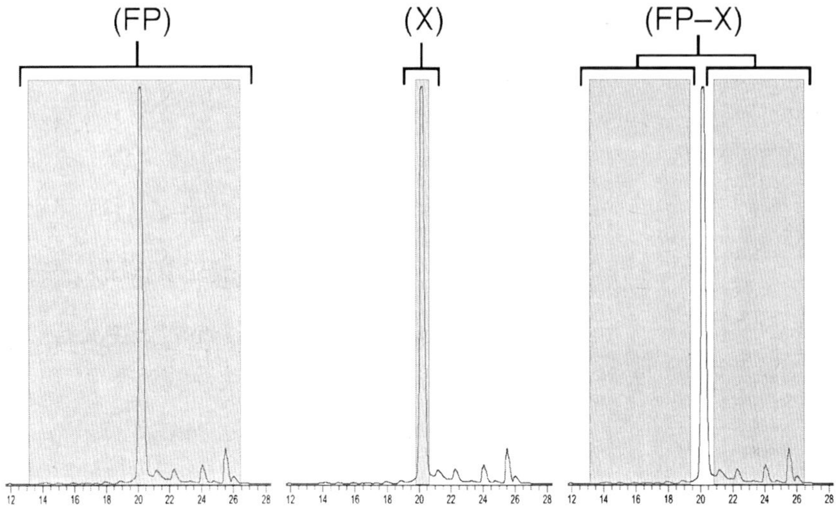

*Figure 2. Fractions collected from kumquat juice by preparative HPLC. FP: flavonoid pool; X: phloretin 3',5'-di-C-glucoside **6**; FP–X: minor flavonoids.*

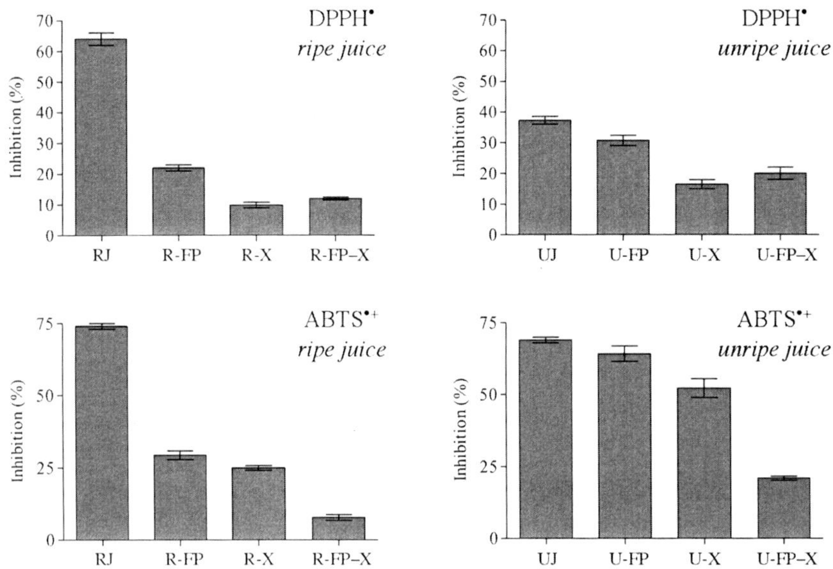

*Figure 3. Antioxidant activity assessed for crude juices and individual fractions obtained from ripe and unripe kumquat fruits. RJ: juice from ripe fruits; UJ: juice from unripe fruits; FP: flavonoid pool; X: phloretin 3',5'-di-C-glucoside **6**; FP–X: minor flavonoids.*

Data in our hands also demonstrate that a much deeper insight into antioxidant properties of complex natural matrices can be obtaining by a separation-oriented antioxidant screening, which may be proposed as an alternative of more common activity-oriented separations. This approach provides interesting evidence on the relative contribution of individual compounds (or selected mixtures thereof) to the total antioxidant capacity of a given food.

Cherimoya (*Annona cherimola* Mill.)

Plants belonging to the *Annona* genus are widely diffused in the tropical and subtropical areas. To date, 119 species have been classified within this genus, but only five of them are currently harvested, and of these, only three – Cherimoya (*A. cherimola*), soursop (*A. muricata*, also known as 'graviola' or 'guanabana') and sugar apple (*A. squamosa*) – are starting to attract the attention of consumers outside the areas where they have been grown as indigenous plants. Among the *Annona* species, cherimoya is the one that has best adapted to subtropical climates or tropical highland areas. It grows as shrub or small tree (5–7.5 m tall), and it has deciduous fruits with oval and slightly oblate shape (10–20 cm long, 7–10 cm wide), with a white creamy pulp containing several dark brown seeds.

During our studies, we have investigated the phenolics content of cherimoya by subjecting methanol, ethanol and dimethylformamide (DMF) extracts of lyophilized pulp to reverse phase HPLC-DAD-ESI-MS-MS analysis (*27*). The chromatographic separation of the three extract presented a chromatogram in which all the detected compounds possess the same UV-Vis spectrum, with an absorption band centered at 278–280 nm, similar to the typical spectrum of flavanone skeletons. Their retention times, UV spectra, MS and MS-MS data and, where available, comparison with standard samples and literature data (*28*), led to the conclusion that the compounds under investigation belonged to the procyanidin group, a subclass of the broader proanthocyanidin family (Table III).

Proanthocyanidins are condensed tannins, and they are the second most abundant natural phenolic after lignin. They are oligo/polymers of flavan-3-ols, where the monomers linked mainly via C4→C8 bonds (in some occasions via the C4→C6 positions). These connections are termed B-type linkages, whereas A-type linkages consist B-type linkages with an additional ether bond between C2→C7 positions. The proanthocyanidins that exclusively consist of (epi)catechin units are designated procyanidins, and they are the most abundant type of proanthocyanidins in plants.

As shown in Table III (*27*), the compounds detected in cherimoya pulp extracts were all procyanidins monomer, dimers and trimers.

The procyanidin dimers (*m/z* 577 $[M-H]^-$, 579 $[M+H]^+$) are characterized by B-types linkage and the MS^2 fragmentation pattern in negative ion mode of of the peak centered at 577 *m/z* gave rise to diagnostic daughter ions at *m/z* 435, 425, 407 and 289. The latter ion is due to quinone methide cleavage of the interflavan bond, and indicates that the procyanidin dimers are (epi)catechin-(epi)catechin. Also the procyanidin trimer are characterized by B-types linkage with a MS^2 spectrum composed of ions at *m/z* 865, 713 [retro-Diels-Alder reaction (RDA)], 577 and

289 (quinone methide cleavage), that allowed us to identify the compounds as (epi)catechin-(epi)catechin-(epi)catechin.

Proanthocyanidins are an interesting class of flavonoids ubiquitously found in some fruits (such as plum and apple, as an example), as well as in tea and red wine. They possess versatile biological effects such as protection against cardiovascular diseases, along with anticancer, antiallergy and antioxidant activities (*29*). As a preliminary screening, the total phenols, total flavonoids and total proanthocyanidins were determined for the three extracts (Table IV) (*27*). Given the high content of potentially antioxidant compounds, the three extracts were subjected to a series of assays.

The extracts obtained from cherimoya pulp were found to be powerful free radical scavengers, and the results of the antioxidant capacity assays on the three extracts towards DPPH• are depicted in Figure 4. All the extracts analyzed showed a remarkable radical scavenging activity towards DPPH•, with the DMF extract (15.3 μM TE) being significantly more efficient than the other two (ethanol: 11.2 μM TE; methanol: 11.6 μM TE).

Table III. Procyanidins Identified in *A. cherimola*

Peak	t_R*(min)*	*Compound*	*[M–H]⁻*	*[M+H]⁺*
1	10.70	(epi)catechin-(epi)catechin	577	579
2	11.89	(epi)catechin-(epi)catechin	577	579
3	12.72	(epi)catechin-(epi)catechin-(epi)catechin	865	867
4	13.88	(epi)catechin-(epi)catechin	577	579
5	15.23	(epi)catechin-(epi)catechin-(epi)catechin	865	867
6	16.03	(epi)catechin-(epi)catechin-(epi)catechin	865	867
7	17.54	(epi)catechin-(epi)catechin	577	579
8	21.71	(epi)catechin-(epi)catechin-(epi)catechin	865	867
9	22.00	(epi)catechin-(epi)catechin-(epi)catechin	865	867
10	22.73	epicatechin	289	291
11	32.28	(epi)catechin-(epi)catechin-(epi)catechin	865	867

Table IV. Total Phenols (TP), Flavonoids (TF) and Proanthocyanidins (TPr) in *A. cherimola* Extracts

Extracts	*TP (μM GA)*[a]	*TF (μM EC)*[b]	*TPr (μM EC)*[b]
Methanol	5.79 ± 0.54	3.51 ± 0.31	1.51 ± 0.12
Ethanol	6.83 ± 0.42	5.72 ± 0.26	4.31 ± 0.35
DMF	6.75 ± 0.83	5.43 ± 0.24	4.91 ± 0.29

[a] Gallic Acid Equivalents. [b] Catechin Equivalents.

Further investigations were carried out by employing the Ferric Reducing Antioxidant Power (FRAP) assay, $ABTS^{\bullet+}$ quenching and Superoxide Radical-anion ($O_2^{\bullet-}$) Scavenging Activity (SRSA). The results of FRAP yielded 1.92±0.074, 1.77±0.188, and 2.75±0.081 μM TE for the methanol, ethanol and DMF extracts, respectively, while results from the $ABTS^{\bullet+}$ and SRSA assays are shown in Figure 4.

Similarly to the $DPPH^{\bullet}$ quenching, also for the $ABTS^{\bullet+}$ and $O_2^{\bullet-}$ assays the decreasing order of efficiency DMF extract > methanol extract > ethanol extract was observed (37.63±2.62, 30.47±0.86 and 22.94±2.2, μM TE for the former assay, and 9.05±0.43, 6.04±0.46 and 5.67±0.38, μM TE for the latter, respectively).

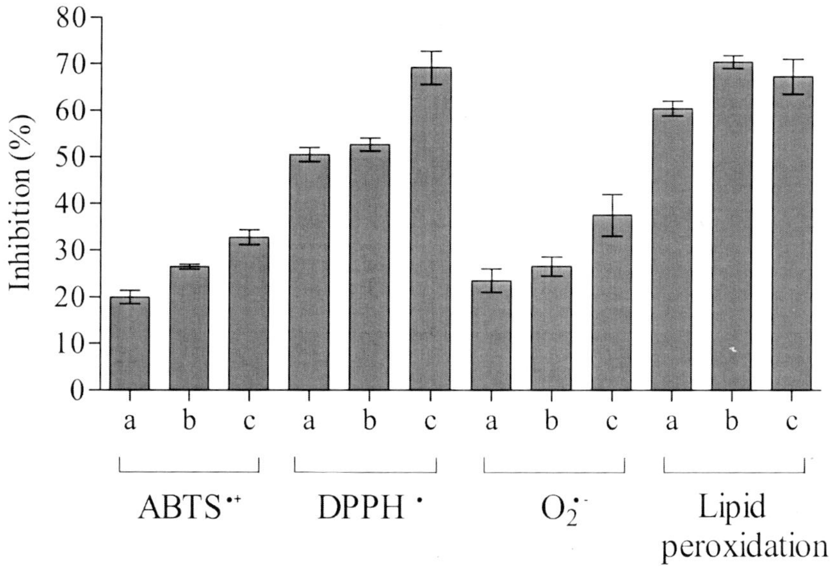

Figure 4. Antioxidant activity assessed for organic extracts of cherimoya pulp by the DPPH•, ABTS•+, SRAS and lipid peroxidation assays. (a) methanol extract; (b) ethanol extract; (c) DMF extract.

Methanol and DMF extracts showed a positive correlation between phenolic content and antioxidant activity. In addition, a very good correlation exists between FRAP–ABTS$^{\bullet+}$, FRAP–DPPH$^{\bullet}$ and DPPH$^{\bullet}$–ABTS$^{\bullet+}$ (*27*), indicating the ability of these extract to scavenge, albeit with varying degree of efficiency, different radicals, such as neutral, cationic and anionic ones.

Data in our hands revealed that the extracts obtained from cherimoya pulp are powerful free radical scavengers and that their primary antioxidant activity may be probably ascribed to the dominant phenolic compounds identified, namely, procyanidins and flavanols (flavan-3-ols). These compounds contain a variety of phenolic hydroxyl groups, and have been shown to possess the strongest antioxidant capacity and free scavenging activity among several phenolic compounds (*30*).

Exposure of organisms to exogenous and endogenous factors usually generates a wide range of reactive oxygen species (ROS), resulting in the homeostatic imbalance and hence negative effects on health (*31*). Therefore, further studies were aimed at the assessment of the lipid peroxidation inhibition and cytoprotective influence of each extracts. The lipid peroxidation assay was carried out by the TBARS method, utilizing erythrocyte membranes and *t*-BuOOH as oxidizing agent. Erythrocyte membranes are among the main subjects of radical attacks in biological systems, as a result of the high concentration of oxygen, hemoglobin and polyunsaturated fatty acids (*32*). Results from the assays of inhibition of lipid peroxidation by methanol, ethanol and DMF cherimoya pulp extracts are shown in Figure 4. 1.0 mM of *t*-BuOOH induced the expected a lipid oxidation, corresponding to 2.77±0.01 μM of malodialdehyde formation in the absence of the extract. Coincubation with ethanol, methanol or DMF extracts markedly decreased the formation of malodialdehyde, with a percentage of inhibition that reached up to 60–70%.

Cherimoya pulp extracts showed also remarkable anticytotoxic properties on human peripheral blood lymphocytes (PBLs) treated with *t*-BuOOH, as evidenced by trypan blue staining. In fact, cell survival decreased responding to treatment with 250 μM *t*-BuOOH for 24 h, with a 34% fall in cell viability compared to control cells. The presence of 0.34 μM GA of ethanol, methanol or DMF extracts significantly improved the cell viability by a factor of 1,34, 1.39 and 1,4, respectively, showing almost the same protective capability (data not shown).

Moreover, interesting information was obtained by monitoring (on the same samples) Caspase 3 activation (commonly utilized as apoptotic marker) and LDH release into the media (often employed as an index of cell membranes integrity). As shown in Figure 5, there was a great increase of LDH release in the *t*-BuOOH treated sample and a statistically significant increase in Caspase 3 activation. Organic extracts, with roughly comparable efficiency were able to completely avoid both events, in agreement with their remarkable antioxidant properties shown by the assays described above.

To further test the antioxidant properties of cherimoya pulp extracts, we have recently carried out experiments on the ability to inhibit protein degradation (Figure 6).

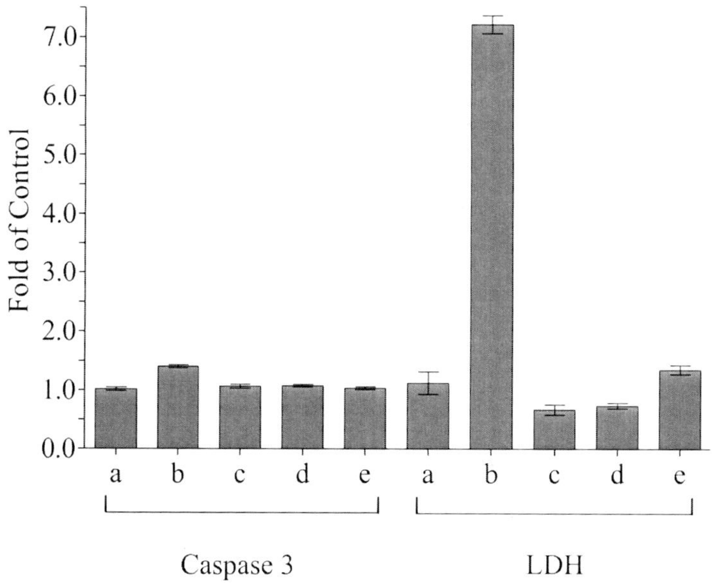

Figure 5. Activation of Caspase 3 and LDH release in t-BOOH treated lymphocytes with or without cherimoya pulp extracts. Samples plus 250 μM of t-BOOH were incubated for 24 h in the absence (b) or in the presence of ethanol (c), methanol (d) and DMF (e) extracts. (a)Samples incubated in the same experimental condition without t-BOOH and pulp extracts.

Human serum albumin (HSA) was employed to analyze the protective effects of the three pulp extracts on protein degradation induced by the Fenton reaction and HOCl attack. The electrophoretic separation of Fenton- or HOCl-treated HSA samples is reported in Figure 6 (*vide infra*). In both cases, in agreement with the data obtained in the antioxidant assays previously described (*27*), the three extracts were able to inhibit protein degradation. After separation and coloration, gels were digitally photographed, and the integrated density of each band was quantified (*33*). The hydroxyl radical, generated *in situ* under these experimental conditions, induced the degradation of ~32% of the protein sample, whereas the coincubation in the presence of methanol, ethanol or DMF extracts, completely inhibited degradation. The same results could be observed in the samples incubated in the presence of HOCl, where, even though protein degradation is clearly more pronounced, the three extracts showed the same efficiency in scavenging the oxidant species, underlining the great ability of the extracts to prevent oxidative injury induced by this agent.

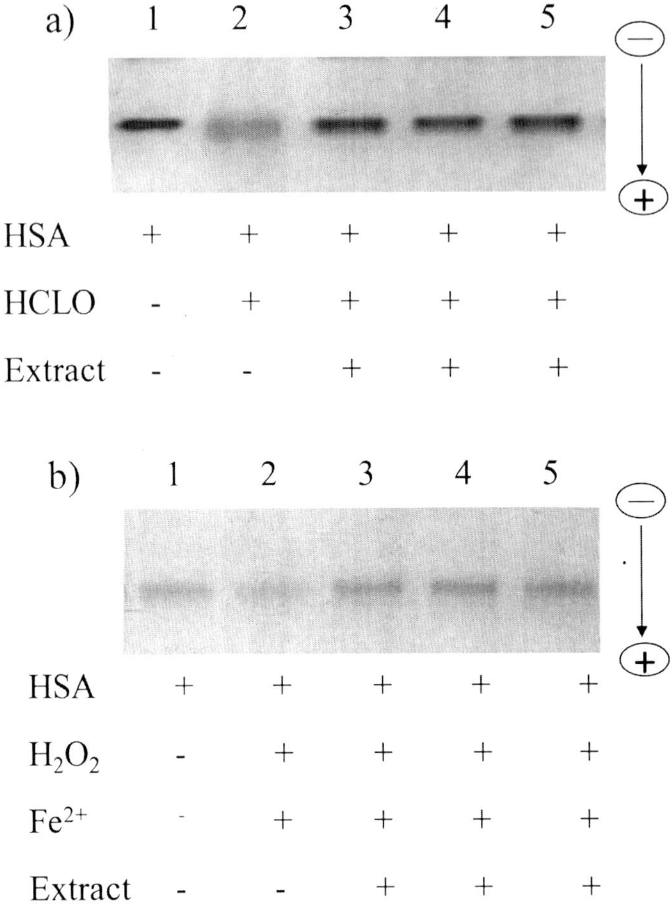

Figure 6. Effect of cherimoya pulp extracts on protein degradation induced by HOCl and Fenton reaction. a) HSA (1 mg/mL) in 20 mM sodium phosphate buffer (pH 7.4) incubated in the absence (1) or in the presence (2) of 1.3% HOCl or the same plus 0.34 μM GA of methanol (3), ethanol (4) or DMF (5) extract. b) HSA (1 mg/mL) in 20 mM sodium phosphate buffer (pH 7.4) incubated in the absence (1) or in the presence (2) of 0.025 mM $FeCl_3$, 0.104 mM EDTA, 0.10 mM ascorbic acid and 2.8 mM H_2O_2, or the same plus 0.34 μM GA of methanol (3), ethanol (4) or DMF (5) extract. Samples were incubated for 40 min at 37 °C and then analyzed by 7.5% polyacrylamide-gel electrophoresis.

Experimental Section

HPLC-DAD-ESI-MS-MS Separation and Identification of Flavonoids

Separation, identification and quantification of the flavonoid profile of kumquat juice and cherimoya pulp has been already reported (*10*, *27*).

Sample Preparation for the Antioxidant and Cytoprotective Assays

Kumquat juice was prepared by hand-squeezing the fruits, after the careful removal of the peel. It was then centrifuged at 4000 rpm for 10 min and the supernatant was collected for analysis, whereas fraction collected from preparative HPLC were evaporated to dryness and redissolved in 2 mL H_2O/DMF 1:1 v/v.

Cherimoya pulp, denied of seed and peel was lyophilized, and three different samples (2 g each) were extracted at room temperature under continuous stirring for 12 h, with methanol, ethanol and DMF, respectively. The samples were then immediately centrifugated at 4000 rpm for 15 min and the supernatants were filtered with filter paper and evaporated to dryness in a rotavapor. These operations were repeated 3 times and the obtained powders were resuspended with the appropriate solvent up to 2 mL and used for the antioxidant and cytoprotective assays.

DPPH, ABTS, superoxide anion, FRAP and inhibition of lipid peroxidation assays were carried out under the conditions we published previously (*10*, *27*).

Cytotoxicity Assays

Human peripheral blood lymphocytes (1×10^6/mL), isolated as previous described (*27*), were incubated in complete medium with or without 0.34 μM gallic acid equivalents (GA) of cheimoya pulp extracts for 24 h in the presence of 250 μM *tert*-butyl hydroperoxide (t-BOOH). In all experiments, parallel controls were performed without t-BOOH. After the incubation period, cell viability has been assessed by trypan blue staining. Citotoxicity was also measured by lactate dehydrogenase (LDH) release from damaged cells into culture medium utilizing a commercially available kit from BioSystems S.A.

For caspase activity determination, cells were collected and washed after the treatment, as previous described (*34*, *35*), to obtain a partial purification of the enzyme. The enzymatic solution was incubated with enzyme-specific colorimetric substrates (Ac-DEVD-pNA) at 37 °C for 1 h. The activity of caspase-3 was analyzed with a spectrophotometer following the release of p-nitroaniline (Pna) at 405 nm and expressed in arbitrary units as a function of untreated sample (*36*).

Inhibition of Protein Degradation

The ability of Cherimoya pulp extracts to inhibit protein degradation induced by HOCl and Fenton reaction was determined utilizing human serum albumin (HSA). In the former experiment, HSA (1 mg/mL) in 20 mM sodium phosphate buffer (pH 7.4) was incubated in the absence or in the presence of 1.3% HOCl or the same plus 0.34 μM GA of methanol, ethanol or DMF. In the latter experiment, HSA (1 mg/mL) in 20 mM sodium phosphate buffer (pH 7.4) was incubated in the absence or in the presence of 0.025 mM $FeCl_3$, 0.104 mM EDTA, 0.10 mM ascorbic acid and 2.8 mM H_2O_2, or the same plus 0.34 μM GA of methanol, ethanol or DMF extract. After 40 min of incubation at 37 °C, both samples were analyzed by 7.5% polyacrylamide-gel electrophoresis.

Conclusions

The plant kingdom is an unending source of bioactive compounds. Researchers worldwide have started, over the past few years, to address in a systematic fashion the study of the composition and properties of so-far little-known or neglected species. In this respect, the largest pool of fruits and vegetables with potential health-benefits can be found in the tropical and subtropical regions.

Here, two examples have been studied. Kumquat (*C. japonica*) and cherimoya (*A. cherimola*) have both been shown to contain a large amount of different phenolics compounds. These compounds (a dihydrochalcone, flavones and flavanones in the case of kumquat, procyanidins in the case of cherimoya) can be safely held responsible for a number of properties, from the remarkable antioxidant activity, to the protein degradation inhibition, down to the cytoprotective action. These studies demonstrate that there is ample room for investigation in the area of tropical fruits properties.

References

1. Scalbert, A.; Johnson, I. T.; Saltmarsh, M. *Am. J. Clin. Nutr.* **2005**, *81*, 215S–217S.
2. Patil, B. S.; Jayaprakasha, G. K.; Murthy, K. N. C.; Vikram, A. *J. Agric. Food Chem.* **2009**, *57*, 8142–8160.
3. Bellocco, E.; Barreca, D.; Laganà, G.; Leuzzi, U.; Tellone, E.; Kotyk, A.; Galtieri, A. *Mol. Cell. Biochem.* **2009**, *321*, 165–171.
4. Fiorani, M.; De Sanctis, R.; Menghinello, P.; Cucchiarini, L.; Cellini, B.; Dachà, M. *Free Radical Res.* **2001**, *34*, 639–648.
5. Harborne, J. B. In *The Flavonoids*; Chapman & Hall; CRC: Boca Raton, FL, 1999.
6. Liu, Z.-Q. *Chem. Rev.* **2010**, *110*, 5675–5691.
7. Robards, K.; Li, X.; Antolovich, M.; Boyd, S. *J. Sci. Food Agric.* **1997**, *75*, 87–101.
8. Zhang, D.-X.; Mabberley, D. J. Citrus. In *Flora of China*; Wu, Z. Y., Raven, P. H., Hong, D. Y., Eds.; Science Press: Beijing, 2008, vol. 11, pp 51–97.
9. Bayer, R. J.; Mabberley, D. J.; Morton, C.; Miller, C. H.; Sharma, I. K.; Pfeil, B. E.; Rich, S.; Hitchcock, R.; Sykes, S. *Am. J. Bot.* **2009**, *96*, 668–685.
10. Barreca, D.; Bellocco, E.; Caristi, C.; Leuzzi, U.; Gattuso, G. *Food. Res. Int.* **2011**, *44*, 2190–2197.
11. Caristi, C.; Bellocco, E.; Panzera, V.; Toscano, G.; Vadalà, R.; Leuzzi, U. *J. Agric. Food Chem.* **2003**, *51*, 3528–3534.
12. Caristi, C.; Bellocco, E.; Gargiulli, C.; Toscano, G.; Leuzzi, U. *Food. Chem.* **2006**, *95*, 431–437.
13. Cuykens, F.; Claeys, M. *J. Mass Spectrom.* **2004**, *39*, 1–15.
14. Gattuso, G.; Barreca, D.; Gargiulli, C.; Leuzzi; Caristi, C. *Molecules* **2007**, *12*, 1641–1673.
15. Nogata, Y.; Sakamoto, K.; Shiratsuchi, H.; Ishii, T.; Yano, M.; Ohta, H. *Biosci. Biotechnol. Biochem.* **2006**, *70*, 178–192.

16. Kawaii, S.; Tomono, Y.; Katase, E.; Ogawa, K.; Yano, M. *J. Agric. Food Chem.* **1999**, *47*, 128–135.
17. Leuzzi, U.; Caristi, C.; Panzera, V.; Licandro, G. *J. Agric. Food Chem.* **2000**, *48*, 5501–5506.
18. Barreca, D.; Bellocco, E.; Caristi, C.; Leuzzi, U.; Gattuso, G. *Food Chem.* **2011**, *129*, 417–422.
19. Kawaii, S.; Tomono, Y.; Katase, E.; Ogawa, K.; Yano, M. *J. Agric. Food Chem.* **1999**, *47*, 3565–3571.
20. Gattuso, G.; Caristi, C.; Gargiulli, C.; Bellocco, E.; Toscano, G.; Leuzzi, U. *J. Agric. Food Chem.* **2006**, *54*, 3929–3935.
21. Gattuso, G.; Barreca, D.; Caristi, C.; Gargiulli, C.; Leuzzi, U. *J. Agric. Food Chem.* **2007**, *55*, 9921–9927.
22. Barreca, D.; Bellocco, E.; Caristi, C.; Leuzzi, U.; Gattuso, G. *Food Chem.* **2011**, *124*, 576–582.
23. Barreca, D.; Bellocco, E.; Caristi, C.; Leuzzi, U.; Gattuso, G. *J. Agric. Food Chem.* **2010**, *58*, 3031–3036.
24. Barreca, D.; Bellocco, E.; Caristi, C.; Leuzzi, U.; Gattuso, G. *Food Chem.* **2011**, *129*, 1504–1512.
25. Wang, Y.-C.; Chuang, Y.-C.; Ku, Y.-H. *Food Chem.* **2007**, *102*, 1163–1171.
26. Huang, D.; Ou, B.; Prior, R. L. *J. Agric. Food Chem.* **2005**, *53*, 1841–1856.
27. Barreca, D.; Laganà, G.; Ficarra, S.; Tellone, E.; Leuzzi, U.; Galtieri, A.; Bellocco, E. *Food Res. Int.* **2011**, *44*, 2302–2310.
28. Huang, W.-Y.; Cai, Y.-Z.; Corke, H.; Sun, M. *J. Food Comp. Anal.* **2010**, *23*, 510–517.
29. Rasmussen, S. E.; Frederiksen, H.; Krogholm, K. S.; Poulsen, L. *Mol. Nutr. Food Res.* **2005**, *49*, 159–174.
30. Cai, Y. Z.; Sun, M.; Xing, J.; Luo, Q.; Corke, H. *Life Sci.* **2006**, *78*, 2872–2888.
31. Halliwell, B.; Gutteridge, J. M. C. *Free Radicals in Biology and Medicine*, 3rd ed.; Oxford University Press: New York, 1999.
32. Barreca, D.; Laganà, G.; Tellone, E.; Ficarra, S.; Leuzzi, U.; Galtieri, A.; Bellocco, E. *J. Membr. Biol.* **2009**, *230*, 163–171.
33. *ImageJ*, version 1.45, NIH. http://rsbweb.nih.gov/ij/index.html (accessed April 23, 2013).
34. Tellone, E.; Ficarra, S.; Russo, A.; Bellocco, E.; Barreca, D.; Laganà, G.; Leuzzi, U.; Pirolli, D.; De Rosa, M. C.; Giardina, B.; Galtieri, A. *Biochimie* **2012**, *94*, 393–402.
35. Ficarra, S.; Tellone, E.; Giardina, B.; Scatena, R.; Russo, A.; Misiti, F.; Clementi, M. E.; Colucci, D.; Bellocco, E.; Laganà, G.; Barreca, D.; Galtieri, A. *J. Membrane Biol.* **2009**, *228*, 43–49.
36. Galtieri, A.; Tellone, E.; Ficarra, S.; Russo, A.; Bellocco, E.; Barreca, D.; Scatena, R.; Laganà, G.; Leuzzi, U.; Giardina, B. *Biol. Chem.* **2010**, *391*, 1057–1065.

Chapter 7

Anti-Inflammatory Mediated Applications of Monoterpenes Found in Fruits

Kotamballi N. Chidambara Murthy,[1] G. K. Jayaprakasha, and Bhimanagouda S. Patil*

Vegetable and Fruit Improvement Center, Department of Horticultural Sciences, Texas A&M University, College Station, Texas 77845-2119, USA
[1]Current address: Division of Research and Patents, M.S.Ramaiah Institute of Health and Allied Sciences, Gokula Education Foundation (Medical), MSRIT Post, New BEL Road, Bangalore-560 054, India
***E-mail: b-patil@tamu.edu. Phone: 979-458-8090. Fax: 979-862-4522.**

Traditionally plants have been used as medicine and food for providing key health benefits, including prevention of a number of infectious, chronic and acute diseases. One of the advantages of plant-derived compounds is their higher margin of safety and their multitude of benefits in prevention of chronic disease and associated symptoms; this activity is either due to isomeric forms of one molecule or a mixture of several molecules, as in extracts and poly herbal formulations. Monoterpenes are abundant volatile principles distributed widely in the plant kingdom and are responsible for different flavors and fragrances. Among monoterpenes, d-limonene is commonly found in citrus and other plants. D-limonene is known for various biological activities, including anti-inflammatory, anticancer, antioxidant and anti-atherosclerotic activities. With the emerging understanding of the association between inflammation and cancer, respiratory diseases and cardiovascular disease, compounds capable of suppressing inflammation are gaining significance as prophylactic and therapeutic agents. The current chapter provides information about various naturally-occurring anti-inflammatory agents, including volatile principles from citrus and the specific benefits of d-limonene in prevention and treatment of cancer and other inflammation-related diseases.

© 2013 American Chemical Society

Keywords: Citrus; volatile oil; phytochemicals; fruits; inflammation; cancer

Introduction

Inflammation is an ancient word originating from Latin word 'inflammare', which means 'to set on fire' and it was used to describe symptoms associated with accidents/injuries. Celsus first documented inflammation in the 1st century AD to describe cardiac signs. Later in 1871, Virchow explained the role of inflammation in cell multiplication for the first time, which was closely associated with cancer development, in 19th century (*1*). The first biochemical definition for inflammation was based on writings by Lewis in 1927, who said that inflammation is a triple response to injury. In general, inflammation is a mechanical/biological event that causes the release of certain chemicals like cytokines, leading to irritation, discomfort or pain in a localized area. Recently, modern molecular biology has shown that inflammation involves a complex cascade of signaling and regulatory proteins. Whenever there is external stimuli of inflammation, the IκB-α (NFκB) protein will be subjected phosphorylation leading to translocation of p50 and p65 units from cytoplasm to nucleus. Upon translocation, it activates the expression of gene which results in activation of apoptosis regulators (Fas, Bcl_2), chemokines (IL-8, human monocyte chemo attractant protein-1), cytokines (IL-2, IL-6), receptors, enzymes like cyclooxygenase-2, prostaglandins, which lead to development of inflammation process (*2*). The ability of multiple proteins to directly or indirectly influence inflammation has clearly shown the association of inflammation with a number of chronic diseases (*3*). For example, inflammation is associated with cardiovascular diseases, notably atherosclerosis (*4*) and congestive heart failure (*5*). Interleukins play a vital role in inflammatory pathway mediated cardiovascular complications. In cancer, activation of nuclear factor κB (NF-κB) and tumor necrotic factor-α, are known to play a pivotal role.

Acute inflammation is a short-term response, which usually results in healing without many complications. It is characterized by the exudation of fluids and plasma proteins along with the migration of leukocytes, most notably neutrophils, into the injured area. This acute inflammatory response is believed to be a defense mechanism to kill various infectious agents such as bacteria, virus and parasites, while facilitating wound repair. The most common chemical mediators produced during acute inflammation are bradykinin, histamine, leukotrienes, prostaglandins, the complement system, anaphylotoxins and nitric oxide (*6*). Natural compounds are known to help in prevention of acute inflammation (*7*), (*8*). Cyclooxygenase (Cox) enzymes are responsible for conversion of arachidonic acid to prostaglandin H_2. Increased formation of prostaglandin H_2 is observed under inflammation conditions. Hence inhibition of Cox, especially Cox-2 is of great significance in inhibition of inflammation process. In our laboratory, we have demonstrated that citrus triterpenoids are capable of suppressing inducers of inflammation, such as NF-κB and Cox-2, in cancer induced in animals (*9*). Recently, we have reported the inhibiting NF-κB and Cox-2 in human pancreatic cancer (MDA Panc-28) cells by citrus obacunone (*10*).

In contrast to acute inflammation, chronic inflammation is a prolonged, dys-regulated and maladaptive response that involves active inflammation, tissue destruction and attempts to repair tissues. This involves monocytes, macrophages, lymphocytes and plasma cells and occurs for longer time. It is characterized by the presence of lymphocytes and macrophages, resulting in fibrosis and tissue necrosis. Persistent chronic inflammation increases the occurrence or severity of degenerative diseases such as rheumatoid arthritis, atherosclerosis, heart disease, Alzheimer's disease, asthma, acquired immunodeficiency disorder, cancer, congestive heart failure, diabetes, infections (bacteria, fungi, parasites), gout, inflammatory bowel disease, aging and other neurodegenerative diseases, all of which are associated with an immunopathological syndrome that appears to play a key role in the onset of the condition (*11*). Inflammation involves several cellular signaling pathways, which are also essential for various normal cellular activities (Figure 1). Developments in biochemical research suggest that NF-κB serves as a key player in linking inflammation with cancer; therefore, targeting the inhibition of NF-κB would provide optimum benefits for management of cancer (*12*). The association of inflammation with cancer opened a new understanding of the complications of cancer and also allowed the targeting of new therapies for the disease (*13*), (*14*). For example, natural compounds identified as reducing chronic inflammation include coumarins, terpenes and flavonoids (*15*), (*16*), (*17*).

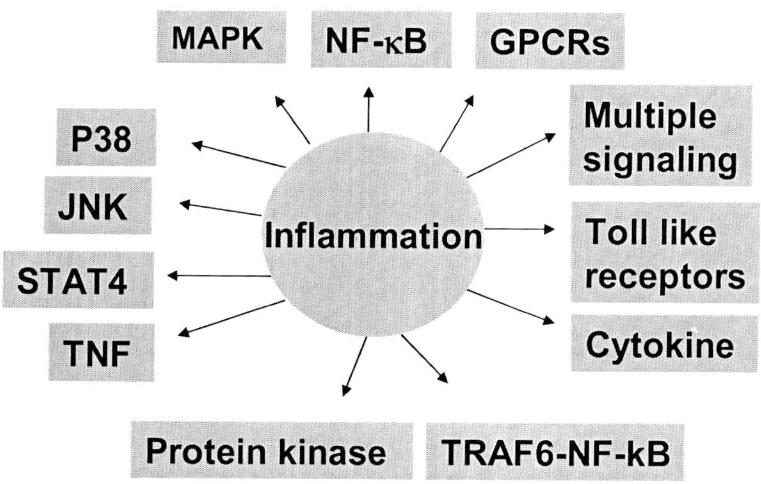

Figure 1. Cellular signaling pathways involved in inflammation. MAPK- Mitogen-activated protein kinase ; NF-κB- nuclear factor κB; JNK- c-Jun N-terminal kinases; STAT4- Signal Transducer and Activator of Transcription; TNF-Tumor necrosis factor; TRAF6- TNF receptor associated factor; GPCRs- G Protein-Coupled Receptors.

Among the possible reasons for spontaneous cancer, two key factors play an important role, exposure to carcinogens and chronic inflammation. Inflammation exclusively accounts for approximately 20% of human cancer and also has a direct or indirect association with most organ-specific cancers. However, the precise mechanistic association of cancer and inflammation is still unclear. It is well known that activation of NF-κB is a key process in inflammation and most commonly associated with cancer (*12*). NF-κB comprises a family of inducible transcription factors that are important in both host defense and inflammation regulation. NF-κB also protects cells from apoptosis and DNA damage under injury or high cytokine conditions (*18*). Hence, compounds capable of inhibiting NF-κB play a vital role in subsidizing the inflammatory reactions and prevention of cancer progression. Some of the pre-clinically and clinically proven natural compounds that can effectively inhibit NF-κB include curcumin, epigallocatechin gallate (EGCG), and resveratrol (*19*).

Monoterpenes in Tropical and Subtropical Fruits and Their Health Promoting Properties

Fruits contain wide range of monoterpenes which have health beneficial effects including prevention of inflammation. Wide ranges of monoterpenes, which have health beneficial effects including prevention of inflammation, are found in fruits. Monoterpenes alcohols namely, linalool, geraniol, nerol, citronellol and α-terpineol are found in annona, citrus, guava, jackfruit, litchi, mango, passion fruits, pineapple and white sapote (Table 1). These compounds are known for their biological activities which includes, antiseptic, antiviral and antifungal activity (*20*). Both α- and β-pinene are known for antibacterial and antifungal and cytotoxicity against murine macrophages. In addition, these compounds have also capable of reducing the bio-film formation by *C. albicans* ATCC10231 and *Aspergillus fumigates* (*21*). Volatile oils rich in germacrene-D has also demonstrated anti-microbial activity against bacteria and fungus (*22*). Volatile oil from some plants rich in germacrene-D reported for antioxidant activity as measured by DPPH radical scavenging assay and inhibition of linoleic acid oxidation (*23*). Another major monoterpene, 1,8-ciniole has been demonstrated repellent, toxicant and protectant against *Sitophilus granarius, S. zeamais, Tribolium castaneum* and *Prostephanus truncatus* (*24*). Apoptosis induction ability of 1, 8-cinole has been reported in Molt 4B, HL-60 and stomach cancer KATO III cells. This was confirmed through induction of DNA fragmentation by these molecules (*25*). Linalool, another major monoterpene was demonstrated anti-inflammatory activity measure using carrageenin-induced edema rats model (*26*). β-Myrecene was also reported for antimicrobial, antioxidant and anti-inflammatory activity *in vitro* and *in vivo* models (*27*), (*28*). Since, D-limonene is one of the most abundant monoterpene in several fruits and we have included comprehensive information about health promoting activity of D-limonene.

Table 1. Monoterpenes Found in Major Tropical and Subtropical Fruits

Fruit	*Major terpenes reported*
Annona atemoya (Custard apple)	α- and β-pinene, E-ocimene, calarene, germacrene-D, 1,8-ciniole, sabinene (*44*), (*45*)
Avocado	Geraniol, limonene, nerol, terpenolene, 1,8-ciniole (*46*)
Breadfruit	O-Geranylsinapic and O-geranylconiferic alcohols (*47*)
Citrus	Limonene, D-dihydrocarvone, β-myrecene, α-pinene (*41*)
Guava	Limonene, myrcene (*48*)
Jackfruit	Ethyl butanolate, ethyl- 3 methyl butanolate, butyl acetate, ethyl- 2 3 methyl butanolate (*49*)
Litchi	Linalool, furaneol (*50*), (*51*)
Mango	α-Pinene, car-3-ene, limonene, γ-terpinene, α-humulene, β-selinene (*52*)
Passion Fruit	Linalool (*53*)
Pineapple	Ocimene (*54*)
White sapote	α-Terpinene (*55*)

Anti-Inflammatory and Anticancer Activities of D-Limonene

D-limonene is the most abundant monoterpene found in citrus including leaf, peel and fruit; indeed, the name "limonene" is derived from lemon. It is also found in dill, cumin, neroli, bergamot, caraway and number of tropical and subtropical fruits widely used (Figure 2). D-limonene is a clear liquid with characteristic lemon odor. It is slightly soluble in water and completely soluble in acetone, dimethyl sulfoxide (DMSO) and ethanol. It has been used as flavor and fragrance for many years, especially in confections and sweet dishes.

D-limonene is known for number of biological activities, which include anti-microbial and antifungal activity against human pathogens (*29*). In addition, this compound is also toxic to number of plant pathogens, including microbes, insects, pests and mites (*30*). Table 2 provides a summary of the various biological activities of d-limonene. It has demonstrated activity in protection against 4-(methylnitrosamino)-1-(3-pyridyl)-1-butanone (NNK) induced lung and fore stomach cancer in female A/J mice. The ability of d-limonene to protect cells from damage was also confirmed by prevention of pulmonary adenoma formation upon administration of d-limonene one hour prior to administration of NNK (*31*). Another study demonstrated that supplementation of 0.5% d-limonene in drinking water one week prior to administration of azoxymethane significantly reduced the development of colon aberrant crypt foci (*32*). This biological effect on microbes serves as the basis for the cytotoxic nature of d-limonene on cultured human cancer cells and implanted tumors (*32*, *33*), (*34*). The most important cancer cell pathways that are affected by d-limonene include induction of phase-I, II enzymes, apoptosis, inhibition of cell cycle progression and anti-inflammatory responses (*35*), (*36*), (*37*).

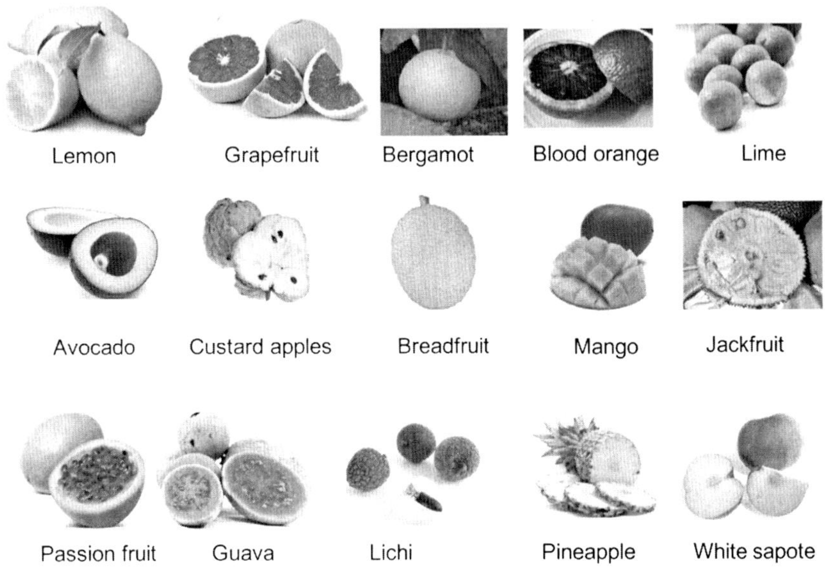

Figure 2. Tropical and subtropical fruits containing d-limonene as one of their active ingredients.

D-limonene has demonstrated antitumor activity against a number of organ-specific tumors in animal models, including mammary, skin, liver, lung and fore stomach. Other monoterpenes found in dietary and non-dietary plants also have demonstrated benefits in prevention of lung, fore stomach and mammary cancers when fed at initial stages of the disease (*38*). The safety and pharmacokinetic behavior of d-limonene was examined in studies with a total of 42 patients, of whom 32 had refractory solid tumors and 10 had breast cancer. The d-limonene was administered to these 32 individuals with refractory solid tumors as 99 courses at a dose of 0.5 to 12 g/m2 per day administered orally in 21-day cycles for individuals with refractory solid tumors and 15 cycles at a dose of 8 g/m2 per day for breast cancer patients. In both cases the compound was well tolerated and there were no symptoms of either adverse reaction or toxicity observed. Furthermore, the peak plasma concentration (Cmax) of d-limonene in these studies ranged from 10.8+/-6.7 to 20.5+/-11.2 microM and the major metabolites observed in serum were perillic acid, dihydroperillic acid, limonene-1,2-diol, uroterpenol and an isomer of perillic acid (*39*). This study concluded that d-limonene has a relatively good margin of safety and can be explored for use in cancer management. The safety of d-limonene for normal cells at the dose that is toxic to cancer cells was determined in a study using a prostate cancer model. This study also demonstrated synergism of cytotoxicity between docetaxel and d-limonene when administered to DU-145 cells (carcinoma) cells when compared to activity on PZ-HPV-7 (Normal prostate epithelial) cells. Generation of ROS was found to be the one of the reasons for induction of apoptosis in cancer cells and the synergistic activity was attributed to an effect on the modulation of proteins associated with the mitochondrial pathway for induction of apoptosis (*40*).

Table 2. Pharmacological and Biological Activities of D-Limonene

Biological activity	*Subject and reference*
Antimicrobial activity	Spectrum of human pathogenic fungi and bacteria (*29*)
Insecticidal, repellent and phytotoxicity	Various organisms including bacteria, fungi, nematodes, mites and insects (*30*)
Immunomodulatory activity	Balb/c mice (*56*)
Anti-*Listeria monocytogenes* activity	*Listeria monocytogenes* serovars 4b and 1/2c (*57*)
Insecticidal	*Rhyzopertha dominica* (F.) and *Tribolium castaneum* (*58*)
Enhance the absorption of drugs	Percutaneous absorption of indomethacin (*59*)
Inhibition of acetylcholinesterase activity	*In vitro* model (*60*)
Suppression of mevalonate pathway	Rodent models (*61*)
Sedative, motor relaxant	Mice (*62*)
Anticarcinogenic	DMBA-induced rat mammary carcinogenesis (*33*)
Anticancer	Nitrosomethylurea-induced mammary tumors (*34*)
Inhibition of polyps and colon cancer	Azoxymethane induced F344 rats model (*63*)

In addition to acting on apoptosis, d-limonene may also inhibit cancer cells through its anti-inflammatory activity. For example, d-limonene did not show significant inhibition of RAW 264.7 macrophage cell viability at to 0.04%; however, at 0.04%, d-limonene exhibited significant inhibition of lipopolysaccharide induced NO production, as measured using the Griess reagent method. In addition, treatment of 0.01, 0.02 and 0.04% d-limonene produced dose-dependent inhibition of PEG2 (prostaglandin E2) suggesting that d-limonene has potential as an anti-inflammatory agent. Furthermore, immunoblotting of cells treated with d-limonene shows down-regulated expression of iNOS and Cox-2. Production of IL-1β, IL-6 and TNF-α was depleted in a dose-dependent manner in cells treated with d-limonene. These results clearly suggest the benefits of d-limonene in prevention of inflammation associated with cancer and other chronic diseases . Patil et al., demonstrated that citrus volatile oil is capable of inhibiting human colon cancer cells through induction of apoptosis (*41*). The anti-inflammatory activity of d-limonene complements its anticancer activity and provides optimal benefits as a prophylactic agent in management of different cancers. Recently, we have demonstrated the proliferation inhibitory activity of human colon cancer cells by volatile components of *Citrus sinensis* (Blood orange). Influence of volatile oil on viability of both SW480 and HT-29 cells are presented in Table 3. These volatile oils are rich in d-limonene are capable of inducing apoptosis in human-derived cancer cells. Additionally, this volatile oil rich in d-limonene has demonstrated inhibition of inflammatory markers including cyclooxygenase-2 (cox-2) and NF-κB in different cancer cells (*42*). Volatile oil from *Citrus limmetioides*, also known as palestine lime was studied to understand the chemical composition and possible effect on proliferation of colon cancer cells. Unlike most of the citrus species, volatile oil of palestine lime contain only 54% of d-limonene. The minor volatiles identified as triacontane, (+)-sabinene, β-myrcene, α-farnesene, R)-(+)-citronellol, (R)-(+)-citronellal by GC-MS analysis. Anticancer potencial of these volatile oil on human colon cancer results suggests that oil can inhibit the proliferation by MTT assay (*43*). Inhibition of 31.6%, 42% and 58.4% was observed at 24 of treatment with 25, 50 and 100 ppm (Table 3), which suggest the potential activity of the components. However, there was no significant increase in the inhibition activity with further increase in time of incubation to 48 and 72 h.

In summary, activation of mediators of inflammation is known to be a major culprit in many different chronic diseases. Hence, targeting the induction of inflammation provides a very promising approach to prevent and treat various ailments including cancer, cardiovascular diseases, metabolic symptoms, respiratory diseases and associated complications. Considering the nature of phytochemicals in dietary ingredients and their optimal activity conditions, it is essential to perform basic research on the individual components for improving human health. For example, d-limonene, a common dietary monoterpene, is known to have great potential in prevention of inflammation under different pathological conditions. Further research is needed on the potency of d-limonene on inflammation associated with different pathological conditions.

Table 3. Influence of Volatile Principles from Blood Orange (*Citrus sinensis*) and Palestain Sweet Lime (*Citrus limmetioides*) Viability of Human Derived Colon Cancer Cells

Volatile oil	Cell line	Concentration (ppm)	% Viability of cell after different incubation time		
			24 h	48 h	72 h
Blood orange (*Citrus sinensis*)	SW480	6.25	83.95 ± 1.35	78.07 ± 2.14	68.21 ± 0.67
		12.5	76.90 ± 3.03	72.84 ± 2.39	56.94 ± 2.29
		25	71.08 ± 3.79	63.60 ± 3.71	48.00 ± 6.02
		50	61.98 ± 3.12	56.78 ± 3.69	44.14 ± 4.08
		75	55.54 ± 3.98	47.69 ± 3.76	34.08 ± 3.24
		100	45.60 ± 3.60	39.92 ± 3.26	25.81 ± 4.37
		200	18.72 ± 2.29	17.14 ± 2.83	9.71 ± 3.79
	HT-29	6.25	94.31 ± 3.66	90.53 ± 1.55	84.06 ± 2.99
		12.5	89.46 ± 4.96	82.62 ± 2.05	79.47 ± 3.97
		25	82.02 ± 2.97	74.79 ± 3.47	72.17 ± 2.54
		50	71.82 ± 3.96	69.48 ± 2.06	67.73 ± 4.33
		75	70.56 ± 3.29	61.49 ± 1.77	59.77 ± 2.99
		100	63.84 ± 3.29	53.83 ± 2.54	46.60 ± 2.18
		200	37.12 ±2.21	28.34 ± 3.44	24.72 ± 6.25
Palestain sweet lime (*Citrus limmetioides*)	SW480	6.25	87.25 ± 2.12	82.66 ± 2.78	81.13 ± 3.45
		12.5	79.16 ±1.22	76.90 ± 2.40	73.61 ± 2.57
		25	68.32 ± 3.76	71.08 ± 1.76	68.00 ± 1.75
		50	58.04 ± 2.45	61.98 ± 1.92	64.14 ± 1.62
		75	52.22 ± 1.60	55.54 ± 1.89	54.08 ± 1.11
		100	41.62 ± 1.65	45.60 ± 0.77	42.47 ± 0.78
		200	22.67 ± 0.64	18.72 ± 0. 39	16.37 ± 0.94

*Values are mean of three replicates (n=9)

Acknowledgments

These results are based on the work supported by the USDA-NIFA # 2010-34402-20875 "Designing Foods for Health" and State funding – 2012-124801 through Vegetable & Fruit Improvement Center-TX State Appropriation. Support of Mr. Swaroop, Mr. Mahesh and Ms. Deepika Varsha for collection of certain literature is gratefully acknowledged.

References

1. Kulkarni, R. S.; Chidley, H. G.; Pujari, K. H.; Giri, A. P.; Gupta, V. S. *Food Chem.* **2012**, *130*, 58.
2. Gilmore, T. D. *Oncogene* **2006**, *25*, 6680.
3. Scott, A.; Khan, K. M.; Cook, J. L.; Duronio, V. *Br. J. Sports Med.* **2004**, *38*, 248.
4. Libby, P. *Am. J. Clin. Nutr.* **2006**, *83*, 456S.
5. Fearon, W. F.; Fearon, D. T. *Circulation* **2008**, *117*, 2577.

6. Robbins, S. L. *Am. J. Med. Sci.* **1958**, *235*, 118.
7. Yuan, G.; Wahlqvist, M. L.; He, G.; Yang, M.; Li, D. *Asia Pac. J. Clin. Nutr.* **2006**, *15*, 143.
8. Gautam, R.; Jachak, S. M. *Med. Res. Rev.* **2009**, *29*, 767.
9. Vanamala, J.; Leonardi, T.; Patil, B. S.; Taddeo, S. S.; Murphy, M. E.; Pike, L. M.; Chapkin, R. S.; Lupton, J. R.; Turner, N. D. *Carcinogenesis* **2006**, *27*, 1257.
10. Murthy, K. N. C.; Jayaprakasha, G. K.; Patil, B. S. *Toxicol. In Vitro* **2011**.
11. Robb-Smith, A. *J. Clin. Pathol.* **1966**, *19*, 205.
12. Pikarsky, E.; Porat, R. M.; Stein, I.; Abramovitch, R.; Amit, S.; Kasem, S.; Gutkovich-Pyest, E.; Urieli-Shoval, S.; Galun, E.; Ben-Neriah, Y. *Nature* **2004**, *431*, 461.
13. Coussens, L. M.; Werb, Z. *Nature* **2002**, 860.
14. Clevers, H. *Cell* **2004**, *118*, 671.
15. Hoult, J.; Paya, M. *Gen. Pharmacol.* **1996**, *27*, 713.
16. Recio, M.; Giner, R.; Uriburu, L.; Manez, S.; Cerda, M.; De la Fuente, J.; Rios, J. *Life Sci.* **2000**, *66*, 2509.
17. Kumazawa, Y.; Kawaguchi, K.; Takimoto, H. *Curr. Pharmaceut. Des.* **2006**, *12*, 4271.
18. Yamamoto, Y.; Gaynor, R. B. *J. Clin. Invest.* **2001**, *107*, 135.
19. Surh, Y.-J.; Chun, K.-S.; Cha, H.-H.; Han, S. S.; Keum, Y.-S.; Park, K.-K.; Lee, S. S. *Mutat. Res., Fundam. Mol. Mech. Mutagen.* **2001**, *480–481*, 243.
20. Kotan, R.; Kordali, S.; Cakir, A. *Z. Naturforsch., C: J. Biosci.* **2007**, *62*, 507.
21. Silva, A. C. R.; Lopes, P. M.; Azevedo, M. M. B.; Costa, D. C. M.; Alviano, C. S.; Alviano, D. S. *Molecules* **2012**, *17*, 6305.
22. Azaz, D.; Demirci, F.; Satil, F.; Kurkcuoglu, M.; Baser, K. *Z. Naturforsch., C: J. Biosci.* **2002**, *57*, 817.
23. Şahin, F.; Güllüce, M.; Daferera, D.; Sökmen, A.; Sökmen, M.; Polissiou, M.; Agar, G.; Özer, H. *Food Control* **2004**, *15*, 549.
24. Obeng-Ofori, D.; Reichmuth, C. H.; Bekele, J.; Hassanali, A. *J. Appl. Entomol.* **1997**, *121*, 237.
25. Moteki, H.; Hibasami, H.; Yamada, Y.; Katsuzaki, H.; Imai, K.; Komiya, T. *Oncol. Rep.* **2002**, *9*, 757.
26. Peana, A. T.; D'Aquila, P. S.; Panin, F.; Serra, G.; Pippia, P.; Moretti, M. D. L. *Phytomedicine* **2002**, *9*, 721.
27. Ciftci, O.; Ozdemir, I.; Tanyildizi, S.; Yildiz, S.; Oguzturk, H. *Toxicol. Ind. Health* **2011**, *27*, 447.
28. Arras, G.; Usai, M. *J. Food Prot.* **2001**, *64*, 1025.
29. Aggarwal, K.; Khanuja, S.; Ahmad, A.; Kumar, T. S.; Gupta, V. K.; Kumar, S. *Flavour Fragrance J.* **2002**, *17*, 59.
30. Ibrahim, M. A.; Kainulainen, P.; Aflatuni, A.; Tiilikkala, K.; Holopainen, J. K. *Agric. Food Sci. Finl.* **2001**, *10*, 243.
31. Wattenberg, L.; Coccia, J. *Carcinogenesis* **1991**, *12*, 115.
32. Kawamori, T.; Tanaka, T.; Hirose, Y.; Ohnishi, M.; Mori, H. *Carcinogenesis* **1996**, *17*, 369.
33. Elson, C. E.; Maltzman, T. H.; Boston, J. L.; Tanner, M. A.; Gould, M. N. *Carcinogenesis* **1988**, *9*, 331.

34. Maltzman, T. H.; Hurt, L. M.; Elson, C. E.; Tanner, M. A.; Gould, M. N. *Carcinogenesis* **1989**, *10*, 781.
35. Crowell, P. L.; Gould, M. N. *Crit. Rev. Oncog.* **1994**, *5*, 1.
36. Lu, X. G.; Zhan, L. B.; Feng, B. A.; Qu, M. Y.; Yu, L. H.; Xie, J. H. *World J. Gastroenterol.* **2004**, *10*, 2140.
37. Bardon, S.; Foussard, V.; Fournel, S.; Loubat, A. *Cancer Lett.* **2002**, *181*, 187.
38. Crowell, P. L. *J. Nutr.* **1999**, *129*, 775S.
39. Vigushin, D. M.; Poon, G. K.; Boddy, A.; English, J.; Halbert, G. W.; Pagonis, C.; Jarman, M.; Coombes, R. C. *Cancer Chemother. Pharmacol.* **1998**, *42*, 111.
40. Rabi, T.; Bishayee, A. *J. Carcinog.* **2009**, *8*, 9.
41. Patil, J. R.; Jayaprakasha, G. K.; Murthy, K. N. C.; Tichy, S. E.; Chetti, M. B.; Patil, B. S. *Food Chemistry* **2009**, *114*, 1351.
42. Murthy, K. N. C.; Jayaprakasha, G. K.; Patil, B. S. *Life Sci.* **2012**, *91*, 429.
43. Jayaprakasha, G. K.; Murthy, K. N. C.; Patil, B. S. In 42nd ACS National Meeting, Denver, CO, 2011.
44. Bartley, J. *Chromatographia* **1987**, *23*, 129.
45. Fournier, G.; Leboeuf, M.; Cavé, A. *J. Essent. Oil Res.* **1999**, *11*, 131.
46. Bozak, K. R.; O'Keefe, D. P.; Christoffersen, R. E. *Plant Physiol.* **1992**, *100*, 1976.
47. Kukovinets, O.; Zainullin, R.; Kislitsyn, M. *Chem. Nat. Compd.* **2006**, *42*, 1.
48. Macleod, A. J.; Gonzalez de Troconis, N. *Phytochemistry* **1982**, *21*, 1339.
49. Bicas, J. L.; Molina, G.; Dionísio, A. P.; Barros, F. F. C.; Wagner, R.; Maróstica, M. R., Jr; Pastore, G. M. *Food Res. Int.* **2011**, *44*, 1843.
50. Ong, P. K. C.; Acree, T. E. *J. Agric. Food Chem.* **1999**, *47*, 665.
51. Garcia, R.; Alves, E. S. S.; Santos, M. P.; Aquije, G. M. F.; Fernandes, A. A. R.; Santos, R. B.; Ventura, J. A.; Fernandes, P. *Braz. J. Microbiol.* **2008**, *39*, 163.
52. MacLeod, A. J.; de Troconis, N. G. *Phytochemistry* **1982**, *21*, 2523.
53. Winterhalter, P. *J. Agric. Food Chem.* **1990**, *38*, 452.
54. Lamikanra, O.; Richard, O. A. *J. Sci. Food Agric.* **2004**, *84*, 1812.
55. Barakat, D. A. *Aust. J. Basic Appl. Sci.* **2011**, *5*, 693.
56. Raphael, T.; Kuttan, G. *Immunopharmacol. Immunotoxicol.* **2003**, *25*, 285.
57. Mourey, A.; Canillac, N. *Food Control* **2002**, *13*, 289.
58. Prates, H.; Santos, J.; Waquil, J.; Fabris, J.; Oliveira, A.; Foster, J. *J. Stored Prod. Res.* **1998**, *34*, 243.
59. Okabe, H.; Takayama, K.; Ogura, A.; Nagai, T. *Drug Des. Delivery* **1989**, *4*, 313.
60. Miyazawa, M.; Watanabe, H.; Kameoka, H. *J. Agric. Food Chem.* **1997**, *45*, 677.
61. Elson, C. E. *J. Nutr.* **1995**, *125*, 1666S.
62. Gurgel do Vale, T.; Furtado, E. C.; Santos, J.; Viana, G. *Phytomedicine* **2002**, *9*, 709.
63. Kawamori, T.; Tanaka, T.; Hirose, Y.; Obnishi, M.; Mori, H. *Carcinogenesis* **1996**, *17*, 369.

Chapter 8

Garcinol from *Garcinia indica*: Chemistry and Health Beneficial Effects

Wenping Tang,[1] Min-Hsiung Pan,[2] Shengmin Sang,[3] Shiming Li,[1] and Chi-Tang Ho[*,1]

[1]Department of Food Science, Rutgers University, 65 Dudley Road, New Brunswick, New Jersey 08901, USA
[2]Department of Seafood Science, National Kaohsiung Marine University, Kaohsiung, Taiwan
[3]Center for Excellence in Post-Harvest Technologies, North Carolina Agricultural and Technical State University, North Carolina Research Campus, 500 Laureate Way, Kannapolis, North Carolina 28081, USA
[*]E-mail: ho@aesop.rutgers.edu.

Garcinol, a polyisoprenylated benzophenone, is isolated from fruit rinds of *Garcinia indica*, which has been used as an ornament, food ingredient as well as traditional medicine for centuries. With its unusual chemical structure, garcinol has generated considerable research interests in recent years because of its remedial qualities against many human diseases and ailments, mostly attributed to its potent antioxidant capabilities. There are increasing scientific evidences which demonstrate the anti-tumorigenic effectiveness of garcinol, and apoptosis is the commonly suggested pathway. Garcinol has been found to modulate several cell signaling pathways involved in apoptosis and cancer development. The present review summarizes the current knowledge on the various biological activities of garcinol, focusing on its anti-tumorigenic effects, as well as the chemistry aspects of this potential chemopreventive agent.

© 2013 American Chemical Society

Introduction

Garcinia indica, commonly known as kokum and belonging to *Guttiferae* family, is a plant native to certain regions of India (*1*). The genus *Garcinia* includes more than 300 species found in the tropics, especially Asia and Africa, and they have various applications in culinary, pharmaceutical, and industrial fields. *Garcinia indica* extract has been shown to have a plethora of therapeutic effects including antifungal, antioxidant, anti-ageing and anti-diabetic (*2–5*). In the folk Indian medicine system, the fruit rinds and leaves are used to treat inflammatory ailments, rheumatic pain and bowel complaints. The fruit of *Garcinia indica* is brownish or purple about the size of an orange, marbled with yellow, and is crowned by the 4-parted, stalk less stigma (*6*).

Chemical studies have shown that the rind of *Garcinia indica* contains proteins, tannins, pectin, sugars, fat, organic acids like (-)-hydroxycitric acid (HCA), hydroxycitric acid lactone and citric acid; the anthocyanins, cyanidin-3-glucoside and cyanidin-3-sambubioside; and the polyisoprenylated phenolics garcinol and isogarcinol (*7*, *8*), with the highest concentration of anthocyanins (2.4 g/100 g of kokum fruit) (*9*) compared with any other natural sources. It is also found that HCA, the major organic acid in kokum leaves and rinds, is a potential anti-obesity and hypocholesterolemic agent (*10*, *11*). Animal studies and clinical trials demonstrated that HCA could reduce body weight, improve energy metabolism and decrease total cholesterol and LDL levels (*10*, *11*). The mechanism might be through suppressing fatty acid synthesis, lipogenesis, and food intake.

The rinds of *Garcinia indica* contain two polyisoprenylated phenolics, garcinol and its colorless isomer isogarcinol. Garcinol (Figure 1), also known as camboginol, resembles curcurmin in structure (*3*), and has been widely studied for its various pharmacological activities such as antioxidant, anti-inflammatory, anti-bacterial, anti-viral, neuroprotective, anti-ulcer and anti-tumor properties. Garcinol can also modulate various key signaling pathways, which is consistent with its pleiotropic activities.

Chemistry Aspects of Garcinol

Chemistry

Garcinol is the active component of *Garcinia indica*. It has positive reaction with $FeCl_3$ because of its phenolic groups (*12*). It is crystallized out as yellow needles (1.5%) from the hexane extract of the fruit rind. The crystals can be dissolved in many organic solvents such as ethanol, methanol, acetone, dimethyl sulfoxide, acetonitrile, ethyl acetate, chloroform, and hexane, but has very poor aqueous solubility (*13*). The molecular formula and the absorption spectral data indicate that the compound is related to isoxanthochymol and more appropriately, in view of the sign of optical rotation, to Cambogin. The presence of an enolisable 1,3-diketone system in the molecule is confirmed by the formation of two isomeric trimethyl ethers, hydrolysable to single dimethyl ether with dilute alkali solution. Alkali degradation of the methyl ether under stronger conditions (20% ethanolic KOH, reflux) yields veratric acid, indicating the presence of a 3,4-dihydroxybenzoyl unit. The UV spectrum of garcinol suggests that the 1,3-diketone system is conjugated to the 3,4-dihydroxybenzoyl moiety. The IR spectrum of the trimethyl ether shows the presence of a saturated carbonyl group (1727 cm^{-1}) and two α,β-unsaturated carbonyl groups (1668 and 1642 cm^{-1}), accounting for all the oxygen atoms. The NMR spectrum of garcinol in $CDCl_3$ shows the presence of two saturated tertiary methyl groups (two singlets at δ 1.01 ppm and 1.17 ppm) and seven $C{=}C{-}CH_3$ groups (signals at δ 1.54 ppm for two methyls and at 1.60, 1.67, 1.70, 1.74 and 1.84 ppm for one methyl each). It also shows signals for a vinylic methylene (δ 4.38, 2 H, broad singlet) and three other olefinic protons (δ 5.0 m) in addition to three aromatic protons (ABX pattern around δ 6.60 and 6.95) and a hydrogen bonded phenolic hydroxyl at δ 18.0 ppm. The mass spectrum of garcinol is very similar to that of xanthochymol exhibiting major peaks at m/e 602(M^+), 465(M^+ $-C_{10}H_{17}$, base peak), 341 (465-C_9H_{16}) and 137 (Dihydroxybenzoyl). These features clearly indicate that the structure of garcinol is biogenetically derivable from maclurin (2,4,6,3′,4′-pentahydroxy-benzophenone) and five isoprenyl units (*14*).

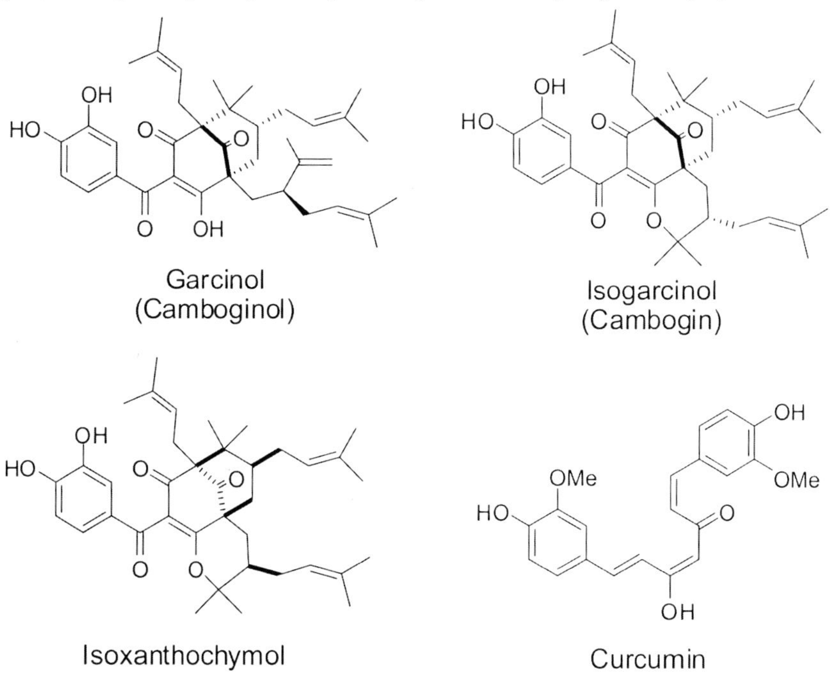

Figure 1. Chemical structures of garcinol, isogarcinol, isoxanthochymol and curcumin.

Treatment of garcinol with acid or heating it to about 200°C yielded a mixture of products, the major product being isogarcinol (Figure 1), and it could be isolated from the mother liquor of garcinol on normal phase chromatography. Its identity with the product from garcinol was established by comparison of UV, IR and NMR spectra (*15*).

Extraction

The procedure for preparation of garcinol from *Garcinia indica* was described in a report in 2000 (*13*). In brief, *Garcinia indica* dried fruit rind was extracted with ethanol. The extract was fractionated by preparative octadecyl silica (ODS) column chromatography eluted stepwise with 70-80% (v/v) ethanol. The eluate was monitored at UV 254 nm, and the main fractions having absorption at 254 nm eluted at 80% (v/v) ethanol were concentrated and dried by rotary evaporator under 50 °C. The dried material was redisolved in hexane and the solution was cooled under 5 °C for 2 days. Yellow amorphous powder was collected from the solution and washed with cold hexane on a glass filter. After drying in a vacuum desiccator, the amorphous substances were solubilized in hot acetonitrile and recrystallized at room temperature; pale yellow crystal needles were obtained from the solvent. The crystals were identified as garcinol. In yet another report *Garcinia indica* rinds were first extracted with water to remove HCA. Then, the dried spent rinds were powdered and successively extracted with hexane and benzene for 4 h each to get 4.4% and 1.8% yields, respectively. Both fractions were mixed in a ratio of 1:1 and loaded onto a silica gel column. The column was eluted with chloroform and methanol with increasing polarity and all fractions were concentrated and analyzed by TLC. Further, fraction 4 was loaded onto Diaion HP-20 column and the column was eluted with water, and mixtures of water and methanol, methanol, and acetone to get 9 fractions. Fraction eluted with methanol gave a single spot on TLC, and it was crystallized to get garcinol (*16*).

Chattopadhyay and Kumar (*17*) developed a liquid chromatography-tandem mass spectrometry (LC-MS/MS) method for the determination of garcinol in the extract of fruit rinds of *Garcinia cambogia*, and they also validated the method in other *Garcinia* species, including *indica*. Separation was achieved isocratically on a reverse phase (RP) C18 column using a solvent system consisting of a mixture of acetonitrile - water (9:1) and methanol - acetic acid (99.5:0.5) in the ratio of 30:70 as mobile phase at a flow rate of 0.4 mL/min and comboginol was quantified using a multiple reaction monitoring (MRM) method. The method showed satisfactory reproducibility with a coefficient of variation of less than 6%; however, the sensitivity of the HPLC method was found to be inadequate for quantifying camboginol and isoxanthochymol in extracts of stem bark, seed and leaves of *Garcinia*. Therefore later the same group developed a more rapid and sensitive liquid chromatography/electrospray ionization tandem mass spectrometrical (LC/ESI-MS/MS) method, with shorter analysis time for LC and LOQ of 4.0 ng/mL for isoxanthochymol and 10 ng/mL for camboginol, which makes the method attractive for the analysis of the above two compounds in different plant parts of *Garcinia* species (*18*). In 2009 Kumar *et al.* reported the intra- and inter-day precisions were 2.34 and 3.41% for isoxanthochymol and 3.35 and 3.66% for camboginol. The identity of the two isomeric compounds in the samples was determined on a triple quadrupole mass spectrometer with ESI interface operating on the negative ion mode (*19*).

Antioxidant Mechanism

Garcinol shows strong antioxidant activity since it contains both phenolic hydroxyl groups as well as a β-diketone moiety, and in this respect it resembles curcumin (Figure 1), a well known antioxidant of plant origin. Antioxidant actions of garcinol are believed to contribute to its chemopreventive activity. Here we summarized the reports on garcinol's potent antioxidant activity and its mechanisms from the literature.

Garcinol was shown to have both chelating activity and free radical scavenging activity and is a lipid-soluble superoxide anion scavenger. The hexane extract, benzene extract and garcinol from *Garcinia indica* have equivalent antioxidant capacity to that of ascorbic acid determined as evaluated by the phosphomolybdenum method and by the 1,1-diphenyl-2-picrylhydrazyl (DPPH) method and among the three garcinol is the highest (*20*). It was shown that under concentrations (1-10 mg/ml) correspondent to the reference of physiological range of tested compounds in plasma, garcinol partially inhibited the effect of peroxynitrite on platelet and plasma lipid peroxidation, measured as thiobarbituric acid reactive substances (TBARS) - a marker of this process, suggesting that garcinol can also be useful as protecting factors against diseases associated with oxidative stress (*20*).

A group from Japan studied the free radical scavenging activity of garcinol in different systems using electron spin resonance (ESR) spectrometry and compared it with a well-known antioxidant DL-R-tocopherol. In the hypoxanthine/xanthine oxidase system, emulsified garcinol suppressed superoxide anion to almost the same extent as DL-R-tocopherol by weight. In the Fenton reaction system, garcinol also suppressed hydroxyl radical more strongly than DL-R-tocopherol. In the H_2O_2/NaOH/DMSO system, garcinol suppressed superoxide anion, hydroxyl radical, and methyl radical. It was thus confirmed that garcinol is a potent free radical scavenger and able to scavenge both hydrophilic and hydrophobic ones including reactive oxygen species (*12*). They later reported that garcinol exhibited moderate antioxidative activity in the micellar linoleic acid peroxidation system (*21*).

In order to provide more insights to the antioxidant mechanism of garcinol, Sang *et al.* characterized the reaction products of garcinol with DPPH (*22*). Two major reaction products, GDPPH-1 and GDPPH-2 (Figure 2) were isolated and identified. Their structures were determined on the basis of detailed high field 1D and 2D NMR spectral analyses. The identification of these products provides the first unambiguous proof that the principle sites of antioxidant reactions are on the 1,3-diketon and the phenolic ring part of garcinol. Garcinol donated an H atom from the hydroxyl group of the enolized 1,3-diketone to form a pair of resonance. All the three compounds were found effective in inducing apoptosis in human leukemia HL-60 cells and in inhibiting NO generation and LPS-induced iNOS gene expression, and GDPPH-1 seem to be more potent than garcinol, with an EC_{50} value of 8.4 μM. In 2002 Sang's group carried out another study to characterize the reaction products of garcinol with peroxyl radicals generated by thermolysis of the azo initiator azo-bisisobutyrylnitrile (AIBN) (*23*). Again with detailed high field 1D and 2D NMR analysis they identified four reaction products and proved that the double bond of the isopentenyl group is a principal site of the antioxidant reaction of garcinol (*51*). Structural elucidation of these products greatly facilitated the understanding of garcinol's potent antioxidant activity.

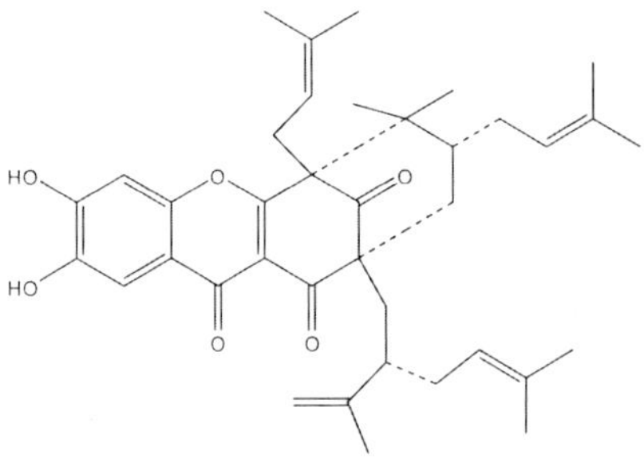

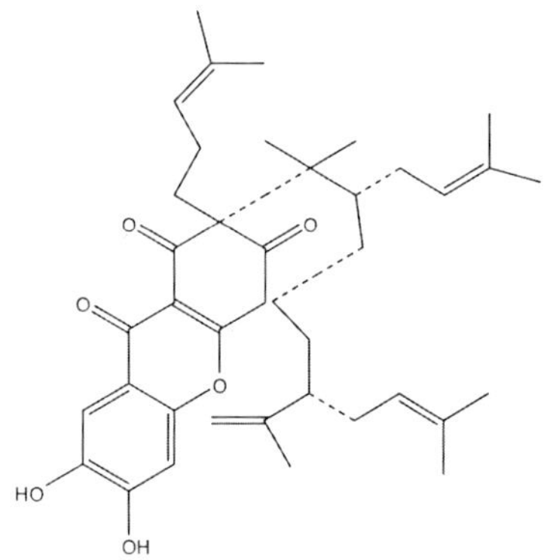

Figure 2. Chemical structures of GDPPH-1 and GDPPH-2.

Biological Activities of Garcinol

Anti-Tumor

The anti-tumorigenic activities of garcinol have been tested both *in vitro* and *in vivo*, and a key mechanism of action for its strong chemopreventive effect is induction of apoptosis. Apoptosis is a form of programmed cell death and it plays a critical role in both development and tissue homeostasis (*24*). Apoptosis is a gene-directed form of cell death which is characterized by cell shrinkage, membrane blebbing, chromatin condensation, and formation of DNA ladder with multiple fragments caused by internucleosomal DNA cleavage (*25*). Proteolytic cleavage of procaspases is an important step leading to caspase activation, which in turn is amplified by the cleavage and activation of other downstream caspases in the apoptosis cascade (*26*, *27*). Caspase-9 is activated when cytochrome c is released into cytoplasm from the mitochondrial intermembranous space. Activated caspase-8 and caspase-9 activate executioner caspases, including caspase-3, which in turn cleave a number of cellular proteins that include structural proteins, nuclear proteins, cytoskeletal proteins, and signaling molecules (*27*).

The mechanism of apoptosis-inducing effect of garcinol was studied in two breast cancer cell lines: estrogen-receptor (ER)-positive MCF-7 and ER-negative MDA-MB-231 cells. Garcinol was found to exhibit dose-dependent cancer cell-specific growth inhibition in both the cell lines with a concomitant induction of caspase-mediated apoptosis through down-regulation of nuclear factor kappa-light-chain-enhancer of activated B cells (NF-κB). Garcinol had no effect on non-tumorigenic MCF-10A cells. These results indicate that garcinol and its derivatives can inhibit intestinal cancer cell growth without affecting normal cells (*28*). Similar apoptosis-inducing effects and mechanism of garcinol was found in prostate (LNCaP, C4-2B and PC3) and pancreatic (BxPC3) cancer cells in a report published recently by the same group. A significant decrease in the colony forming ability of all the cell lines was also observed (*29*). Another group found that Nicotine-induced human breast cancer (MDA-MB-231) cell proliferation was inhibited by 1 μM of garcinol through down-regulation of α9-nAChR and cyclin D3 expression (*30*).

Garcinol resembles curcumin (Figure 1) structurally. In 2001 the effects of garcinol and curcumin on cell viability in human leukemia HL-60 cells were investigated. Both garcinol and curcumin displayed strong growth inhibitory effects against human leukemia HL-60 cells, with estimated IC_{50} values of 9.42 and 19.5 μM, respectively. Garcinol was able to induce apoptosis in a concentration- and time-dependent manner and had a stronger effect than curcumin. The mechanism suggested by the authors is that garcinol induced apoptosis is triggered by the release of cytochrome c into the cytosol, procaspase-9 processing, activation of caspase-3 and caspase-2, degradation of Poly (ADP-ribose) polymerase (PARP), and DNA fragmentation caused by the caspase-activated deoxyribonuclease through the digestion of DNA fragmentation factor-45 (DFF-45) (*31*).

The *in vitro* effects of three benzophenone derivatives, namely garcinol, isogarcinol, and xanthochymol (Figure 1) were examined on cell growth in four human leukemia cell lines including NB4, HL60, U937, and K562. All of the compounds exhibited significant growth suppression in the four cell lines due to apoptosis mediated by the activation of caspase-3. A loss of mitochondrial membrane potential was found in garcinol- and isogarcinol-induced apoptosis, but not in xanthochymol-induced apoptosis. The growth inhibitory effects of isogarcinol and xanthochymol were more potent than that of garcinol (*32*). In another report, two human pancreatic cancer cell lines, BxPC-3 and Panc-1, with wild and mutant k-ras, respectively, were utilized to study the effect of garcinol on pancreatic cancer (PaCa) cell viability and proliferation. Garcinol treatment (0–40 μM) dose- and time-dependently inhibited ($P < 0.05$) cell growth (trypan blue exclusion) by induction of apoptosis via G0-G1 phase cell cycle arrest in both cell lines (*33*).

The beneficial effects of tumor prevention by garcinol on the human colorectal cancer cell line, HT-29 was also explored. Exposure of HT-29 cells to 10 mM garcinol inhibited cell invasion, and decreased the dose dependent tyrosine phosphorylation of FAK thus subsequently down-regulated MAPK/ERK, and PI3K/Akt signaling pathways. At dosage of 20 μM garcinol changed the ratio of the anti-apoptotic Bcl-2 and pro-apoptotic BAX proteins within 12 h, which correlated with a release of cytochrome c from the mitochondria to the cytosol, and with PARP cleavage (*34*).

TRAIL (tumor necrosis factor–related apoptosis-inducing ligand) is a cytokine known to induce apoptosis in a variety of tumor cells. It was found that garcinol can potentiate TRAIL induced apoptosis through up-regulation of death receptor 4 (DR4) and DR5 and down-regulation of anti-apoptotic proteins including survivin, bcl-2, XIAP, and cFLIP; it could also induce bid cleavage, bax, and cytochrome c release in a variety of cancer cells including HCT116 (human colon adenocarcinoma), HT29 (human colon adenocarcinoma), A293 (human embryonic kidney carcinoma), PC3 (human prostate cancer cells), MDA-MB-231 and MCF-7 (human breast cancer cells), U266 (human multiple myeloma), SEG-1 (human esophageal epithelial cells), and KBM-5 (human chronicleukemic cells) (*35*).

In contrast to the above reported inhibitory effects, Yang's group pointed out that garcinol could significantly increase cell proliferation under low concentrations (*9*). They found that garcinol had a biphasic action in modulating intestinal cell growth in vitro: garcinol at <1 μM causes the generation of hydrogen peroxide and stimulates cell proliferation possibly by activating ERK 1/2 and AKT, but at higher concentrations it inhibits cell growth possibly by inducing apoptosis. They suggested that these effects may also be seen in other cell types and the net effects depend on the concentrations of garcinol in different tissues and its metabolites as well as the oxygen partial pressure (*36*).

Studies on anti-tumorigenic effects of garcinol in animal models are less frequently carried out compared to cell culture studies. An earlier report investigated the modifying effects of dietary feeding of garcinol on the development of azoxymethane (AOM)-induced colonic aberrant crypt foci (ACF) in male F344 rats (*37*). The study also assessed the effects of garcinol on proliferating cell nuclear antigen (PCNA) index in ACF and activities of detoxifying enzymes such as glutathione S-transferase (GST) and quinone reductase (QR) in liver. It was observed that garcinol administration both at 0.01% and 0.05% in diet significantly lowers PCNA index in ACF and dramatically elevates liver GST and QR activities. In addition, garcinol was also found to suppress O_2^- and nitric oxide (NO) generation and expression of inducible NO synthase (iNOS) and COX2 proteins (*37*).

Yoshida and coworkers have demonstrated that garcinol effectively inhibits 4-NQO-induced tongue carcinogenesis in male F344 rats without causing any adverse effects. Dietary garcinol (500 ppm in diet) significantly decreased the incidence and multiplicity of 4-NQO-induced tongue neoplasms and/or preneoplasms as compared to the control diet through inhibition of cell proliferation activity suggested from reduced BrdU-labeling index and cyclin D1-positive cell ratio. The COX-2 expression in the tongue lesions was also suppressed by feeding with garcinol (*38*). These results indicate that garcinol might be an agent that exerts cancer chemopreventive ability in oral cavity.

Our group studied the anti-tumor effects of dietary garcinol in prostate and colon cancer animal models. Chemoprevention effect of dietary garcinol was observed both in AOM-induced acute colitis and in AOM/DSS (dextran sulfate sodium) induced colon carcinogenesis in CD-1 mice. Garcinol (250 ppm and 500 ppm in diet) significantly attenuated DSS induced ACFs and dose-dependently prevented shortening of colon length in acute colitis model; in AOM/DSS induced two stage colon tumorigenesis model, the same level of garcinol significantly reduced tumor count and levels of biomarkers on proliferation. In the PC-3 xenorgft prostate cancer mice model, we observed a significant reduction both in the number of tumor counts in the nude mice and also in apoptosis-related protein expressions from garcinol which was administerd intraperitoneally at 50 mg/kg. These results suggest that garcinol may be one of the potential chemotherapy agents of human prostate cancer and colon cancer (unpublished data).

Anti-Inflammation

Garcinol shows great potential as a novel chemopreventive agent, based on its inhibition of lipopolysaccharide (LPS)-induced COX-2 and iNOS expression via regulating signaling pathway including p38 mitogen-activated kinase (MAPK), IκBα, and the p65 subunit translocation of NF-κB (*39*). Topical application or oral administration of garcinol to CD-1 mice markedly inhibited TPA-induced ear inflammation in a dose-dependent manner as well as markedly inhibited TPA-induced up-expression of pro-inflammatory cytokine interluckin-6 (IL-6) protein levels in mouse ears in dose-dependent fashion. Oral administration of garcinol inhibited UVB-induced ear inflammation, and blocked up-expression of pro-inflammatory cytokine IL-1β and IL-6 protein levels in ears, showing that it has good bioavailability (*40*). From studying the effects of garcinol and its derivatives, including cambogin, garcim-1 and garcim-2, on arachidonic acid metabolism and nitric oxide (NO) synthesis in LPS-stimulated RAW264.7 murine macrophages as well as in three intestinal cell lines, the authors found that garcinol modulates arachidonic acid metabolism by inhibiting the activation of phosphorylation of cytosolic phospholipase A2 (cPLA2) through the inhibition of extracellular signal-regulated kinase 1/2 (ERK 1/2) phosphorylation, and suppresses iNOS expression (and NO formation) by inhibiting Janus Kinase (JAK) signal transducer and activator of transcription-1 (STAT-1) activation. When added before LPS, garcinol suppressed NF-κB activation and COX-2 expression through the interruption of LPS binding to toll-like receptors. These inhibitory effects may be the key mechanisms for the anti-inflammatory actions of garcinol (*41*). Another study shows that garcinol potently interferes with the catalytic activity of 5-lipoxygenase (5-LO) and microsomal prostaglandin E_2 (PGE_2) synthase (mPGES)-1, enzymes that play pivotal roles in inflammation and tumorigenesis in cell-free and cell-based assays. Inhibition of mPGES-1 and 5-LO may also be an important pathway for anti-inflammatory effectiveness of garcinol (*42*).

Anti-Ulcer

Hydroxyl radicals can cause various diseases including stress-induced gastric ulcer (*43*) and nonsteroidal anti-inflammatory drug-induced gastric ulcer (*44*), therefore, garcinol is expected to be useful for preventing such diseases since it was found to be a potent radical scavenging agent, especially against hydroxyl radicals (*45*). Actually, orally administered garcinol prevented acute ulceration in rats induced by indomethacin and water immersion stress caused by radical formation (*20*). It was suggested that garcinol might have such potentials in clinical application as an antiulcer drug and the proposed mechanism is that garcinol protects cell from injury by scavenging reactive oxygen species on the surface of gastric mucosa (*21*).

Anti-Microbial

The minimum inhibitory concentration (MIC) of garcinol against Methicillin-Resistant *Staphylococcus aureus* (MRSA) ranges from 6.25-25μg/ml and is equivalent to that of the antibiotic vancomycin which is used to treat MRSA infections (*12*). The MIC of hexane extract, benzene extract of *Garcinia indica*, and garcinol against a few Gram-positive and Gram-negative bacteria were in the range of 15 to 1000, 20 to 1250, and 1.5 to 500 ppm, respectively. Garcinol was effective against *Bacillus sp.* and *S. aureus* at 2 ppm and *L. monocytogenes* at 25 ppm (*16*). Garcinol is much more potent than resveratrol as a bactericidal agent against *Helicobacter pylori* and shows equivalent or better activity compared to clarithromycin at 6 and 12 h incubation (*46*). It was also shown that garcinol was the most effective agent in inhibiting growth of *H. pylori*, among a group of antioxidants including resveratrol, vitamin C, and vitamin E (*47*).

Bakana *et al.* found that garcinol showed promising antifungal activities as compared to an antifungal drug amphotericin B against all fungi and dermatophytes tested, including *Aspergillus flavus, Aspergillus fumigates, Aspergillus niger, Microsporum canis* and *Trichophyton mentagrophytes*, and the MIC values were around 100 μg/ml. But garcinol was inactive in the concentrations used against yeasts tested, nor did it show any antivirus activity in this investigation (*48*).

The low MIC values with garcinol against some of the most important food poisoning and spoilage organisms reveals an exciting potential for its application in food systems. Furthermore, the high antioxidant activity of garcinol will be an added advantage in providing a safe and natural alternative to chemical preservatives.

Anti-HAT

Histone acetylation is a diagnostic feature of transcriptionally active genes. The proper recruitment and function of histone acetyltransferases (HATs) and deacetylases (HDACs) are key regulatory steps for gene expression and cell cycle. Dysfunction of HATs leads to several diseases including cancer, diabetes, and asthma. Recently garcinol was found to be a potent inhibitor of histone acetyltransferases both *in vitro* and in *vivo* in HeLa cells. The microarray analysis of garcinol-treated HeLa cell gene expression also showed that more than 72% of genes tested were down-regulated (*49*). Garcinol was confirmed to be a molecule inhibiting HATs (IC_{50} = 7 μM) and the mechanism is through inducing alteration in the ordered segment of secondary structure of the protein (*50*).

Neuroprotective

The neuroprotective effects of garcinol were examined in a neuro-astrocyte co-culture system where garcinol enhanced neuronal attachment and neurite outgrowth with LPS or IL-1β stimulation. Garcinol also decreased the protein levels of iNOS in LPS-stimulated primary astrocytes, suggesting a neuroprotective effect of garcinol in inflammatory-mediated neurotoxicity (*51*). A recent report provides additional properties of garcinol in neuroprotection and neurite outgrowth of cultured rat cortical progenitor cells for understanding the roles of garcinol in neuronal survival and differentiation. It has been found that the duration of ERK activation contributed to the regulation of neural maturation by garcinol-treatment. Their findings suggested that garcinol enhanced neuronal survival and regulated the potential effects of neuritogenesis (*52*).

Conclusion

In summary, the genus *Garcinia* contains many pharmacologically and medicinally important chemicals entities, with garcinol being the main active principle and a multi-purpose biological agent. Its novel structure makes it a very efficient scavenger of free radicals and an excellent inhibitor of NO. It has been clearly established that the antioxidant nature is the underlying mechanism for many of its various biological activities, and past studies have revealed some of the principal sites for its antioxidant action, although more detailed and in-depth studies are required for establishing the pathway (*53*). It has also been suggested by emerging data from cell culture and animal studies that garcinol has excellent cytotoxicity and anti-tumorigenesis activities, through induction of apoptosis. It has been found that the 1, 2 carbon-carbon double bond of the α, β-unsaturated ketone is important for apoptosis-inducing activity, and garcinol has also been shown to modulate various key signaling pathways, including NF-κB, which is considered very important for its potential role as a chemopreventive agent. In light of the current knowledge on garcinol, we still need more *in vivo* studies in support of the role of garcinol as antitumor agent against human malignancies using suitable higher animal species, and we need to provide the molecular mechanism of action for other health benefits of garcinol which are less well characterized. It will also be of great significance to study the efficacy of garcinol in combination with other herbs or drugs to further explore enhancement and allersteric effects.

References

1. Baliga, M. S.; Bhat, H. P.; Pai, R. J.; Boloor, R.; Palatty, P. L. *Food Res. Internat.* **2011**, *44*, 1790–1799.
2. Joseph, G. S.; Jayaprakasha, G. K.; Selvi, A. T.; Jena, B. S.; Sakariah, K. K. *Int. J. Food Microbiol.* **2005**, *101*, 153–160.
3. Mishra, A.; Bapat, M. M.; Tilak, J. C.; Devasagayam, T. P. A. *Curr. Sci.* **2006**, *91*, 90–93.
4. Sahasrabudhe, A.; Deodhar, M. *Int. J. Bot.* **2011**, *6* (3), 299–303.
5. Kirana, H.; Srinivasan, B. P. *J. Young Pharm.* **2011**, *2*, 265–268.
6. Hemshekhar, M.; Sunitha, K.; Santhosh, M. S.; Devaraja, S.; Kemparaju, K.; Vishwanath, B. S.; Niranjana, S. R.; Girish, K. S. *Phytochem. Rev.* **2011**, *10*, 325–351.
7. Krishnamurthy, N.; Sampathu, S. R. *J. Food Sci. Technol.* **1988**, *25*, 44–45.

8. Nayak, C. A.; Rastogi, N. K.; Raghavarao, K. S. M. S. *Int. J. Food Prop.* **2011**, *13*, 441–453.
9. Nayak, C. A.; Srinivas, P.; Rastogi, N. K. *Food Chem.* **2009**, *118*, 719–724.
10. Heymsfield, S. B.; Allison, D. B.; Vasselli, J. R.; Pietrobelli, A.; Greenfield, D.; Nunez, C. *J. Am. Med. Assoc.* **1998**, *280*, 1596–1600.
11. Mattes, R. D.; Bormann, L. *Physiol. Behav.* **2000**, *71* (1/2), 87–94.
12. Iinuma, M.; Tosa, H.; Tanaka, T.; Kanamaru, S.; Asai, F.; Kobayashi, Y.; Miyauchi, K.-i.; Shimano, R. *Biol. Pharm. Bull.* **1996**, *19*, 311–314.
13. Yamaguchi, F.; Ariga, T.; Yoshimura, Y.; Nakazawa, H. *J. Agric. Food Chem.* **2000**, *48*, 180–185.
14. Rao, A. V. R.; Venkatswamy, G. *Indian J. Chem., Sect. B* **1981**, *20B*, 983–984.
15. Krishnamurthy, N.; Lewis, Y. S.; Ravindranath, B. *Tetrahedron Lett.* **1981**, *22*, 793–796.
16. Negi, P. S.; Jayaprakasha, G. K. *J. Food Sci.* **2004**, *69*, M61–M65.
17. Chattopadhyay, S. K.; Kumar, S. *Biomed. Chromatogr.* **2007**, *21*, 55–66.
18. Chattopadhyay, S. K.; Kumar, S. *Biomed. Chromatogr.* **2007**, *21*, 1159–1165.
19. Kumar, S.; Sharma, S.; Chattopadhyay, S. K. *Biomed. Chromatogr.* **2009**, *23* (8), 888–907.
20. Kolodziejczyk, J.; Masullo, M.; Olas, B.; Piacente, S.; Wachowicz, B. *Platelets* **2009**, *20*, 487–492.
21. Yamaguchi, F.; Saito, M.; Ariga, T.; Yoshimura, Y.; Nakazawa, H. *J. Agric. Food Chem.* **2000**, *48* (6), 2320–2325.
22. Sang, S. M.; Pan, M. H.; Cheng, X. F.; Bai, N. S.; Stark, R. E.; Rosen, R. T.; Lin-Shiau, S. Y.; Lin, J. K.; Ho, C. T. *Tetrahedron* **2001**, *57*, 9931–9938.
23. Sang, S. M.; Liao, C. H.; Pan, M. H.; Rosen, R. T.; Lin-Shiau, S. Y.; Lin, J. K.; Ho, C. T. *Tetrahedron* **2002**, *58*, 10095–10102.
24. Pan, M.-H.; Ghai, G.; Ho, C.-T. *Mol. Nutr. Food Res.* **2008**, *52*, 43–52.
25. Boise, L. H.; Gonzalez-Garcia, M.; Postema, C. E.; Ding, L.; Lindsten, T.; Turka, L. A.; Mao, X.; Nunez, G.; Thompson, C. B. *Cell (Cambridge, Mass.)* **1993**, *74*, 597–608.
26. Earnshaw, W. C.; Martins, L. M.; Kaufmann, S. H. *Annu. Rev. Biochem.* **1999**, *68*, 383–424.
27. Salvesen, G. S.; Dixit, V. M. *Cell (Cambridge, Mass.)* **1997**, *91*, 443–446.
28. Ahmad, A.; Wang, Z.; Ali, R.; Maitah, M. i. Y.; Kong, D.; Banerjee, S.; Padhye, S.; Sarkar, F. H. *J. Cell. Biochem.* **2010**, *109*, 1134–1141.
29. Ahmad, A.; Wang, Z.; Wojewoda, C.; Ali, R.; Kong, D.; Maitah, M. i. Y.; Banerjee, S.; Bao, B.; Padhye, S.; Sarkar, F. H. *Front. Biosci. (Elite Ed)* **2011**, *3*, 1483–1492.
30. Chen, C.-S.; Lee, C.-H.; Hsieh, C.-D.; Ho, C.-T.; Pan, M.-H.; Huang, C.-S.; Tu, S.-H.; Wang, Y.-J.; Chen, L.-C.; Chang, Y.-J.; Wei, P.-L.; Yang, Y.-Y.; Wu, C.-H.; Ho, Y.-S. *Breast Cancer Res. Treat.* **2010**, *125*, 73–87.
31. Pan, M.-H.; Chang, W.-L.; Lin-Shiau, S.-Y.; Ho, C.-T.; Lin, J.-K. *J. Agric. Food Chem.* **2001**, *49*, 1464–1474.
32. Matsumoto, K.; Akao, Y.; Kobayashi, E.; Ito, T.; Ohguchi, K.; Tanaka, T.; Iinuma, M.; Nozawa, Y. *Biol. Pharm. Bull.* **2003**, *26*, 569–571.

33. Parasramka, M. A.; Gupta, S. V. *Nutr. Cancer*, *63*, 456–465.
34. Liao, C.-H.; Sang, S.; Ho, C.-T.; Lin, J.-K. *J. Cell. Biochem.* **2005**, *96*, 155–169.
35. Prasad, S.; Ravindran, J.; Sung, B.; Pandey, M. K.; Aggarwal, B. B. *Mol. Cancer Ther.* **2011**, *9*, 856–868.
36. Hong, J.; Kwon, S. J.; Sang, S. M.; Ju, J. Y.; Zhou, J. N.; Ho, C. T.; Huang, M. T.; Yang, C. S. *Free Radicals Biol. Med.* **2007**, *42*, 1211–1221.
37. Tanaka, T.; Kohno, H.; Shimada, R.; Kagami, S.; Yamaguchi, F.; Kataoka, S.; Ariga, T.; Murakami, A.; Koshimizu, K.; Ohigashi, H. *Carcinogenesis* **2000**, *21*, 1183–1189.
38. Yoshida, K.; Tanaka, T.; Hirose, Y.; Yamaguchi, F.; Kohno, H.; Toida, M.; Hara, A.; Sugie, S.; Shibata, T.; Mori, H. *Cancer Lett. (Amsterdam, Neth.)* **2005**, *221*, 29–39.
39. Liao, C.-H.; Sang, S.; Liang, Y.-C.; Ho, C.-T.; Lin, J.-K. *Mol. Carcinog.* **2004**, *41*, 140–149.
40. Huang, M.-T.; Liu, Y.; Badmaev, V.; Ho, C.-T. Antiinflammatory and Anticancer Activities of Garcinol. In *Dietary Supplements*; Ho, C.-T., Simon, J. E., Shahidi, F., Shao, Y., Eds.; ACS Symposium Series 987; American Chemical Society: Washington, DC, 2008; pp 293–303.
41. Hong, J. G.; Sang, S. M.; Park, H. J.; Kwon, S. J.; Suh, N. J.; Huang, M. T.; Ho, C. T.; Yang, C. S. *Carcinogenesis* **2006**, *27*, 278–286.
42. Koeberle, A.; Northoff, H.; Werz, O. *Biochem. Pharmacol.* **2009**, *77*, 1513–1521.
43. Das, D.; Bandyopadhyay, D.; Banerjee, R. K. *Free Radicals Biol. Med.* **1997**, *24*, 460–469.
44. Vaananen, P. M.; Meddings, J. B.; Wallace, J. L. *Am. J. Physiol.* **1991**, *261* (3) (Pt. 1), G470–G475.
45. Yamaguchi, F.; Saito, M.; Ariga, T.; Yoshimura, Y.; Nakazawa, H. *J. Agric. Food Chem.* **2000**, *48*, 2320–2325.
46. Chatterjee, A.; Yasmin, T.; Bagchi, D.; Stohs, S. J. *Mol. Cell. Biochem.* **2003**, *243*, 29–35.
47. Chatterjee, A.; Bagchi, D.; Yasmin, T.; Stohs, S. J. *Mol. Cell. Biochem.* **2005**, *270*, 125–130.
48. Bakana, P.; Claeys, M.; Tott, J.; Pieters, L. A. C.; Van Hoof, L.; Tamba, V.; Van Den Berghe, D. A.; Vlie-tinck, A. J. *J. Ethnopharmacol.* **1987**, *21*, 75–84.
49. Balasubramanyam, K.; Altaf, M.; Varier, R. A.; Swaminathan, V.; Ravindran, A.; Sadhale, P. P.; Kundu, T. K. *J. Biol. Chem.* **2004**, *279* (32), 33716–33726.
50. Arif, M.; Pradhan, S. K.; R, T. G.; Vedamurthy, B. M.; Agrawal, S.; Dasgupta, D.; Kundu, T. K. *J. Med. Chem.* **2009**, *52*, 267–277.
51. Liao, C. H.; Ho, C. T.; Lin, J. K. *Biochem. Biophys. Res. Commun.* **2005**, *329*, 1306–1314.
52. Weng, M.-S.; Liao, C.-H.; Yu, S.-Y.; Lin, J.-K. *J. Agric. Food Chem.*, *59*, 1031–1040.
53. Padhye, S.; Ahmad, A.; Oswal, N.; Sarkar, F. H. *J. Hematol. Oncol.* **2009**, *2*, 13.

Chapter 9

Beneficial Role of L-Cysteine and H_2S Rich Fruits and Vegetables in Diabetic Pathophysiology

Prasenjit Manna and Sushil K. Jain*

Department of Pediatrics, Louisiana State University, Health Sciences Center, 1501 Kings Highway, Shreveport, Louisiana 71130
***E-mail: sjain@lsuhsc.edu. Tel: 1-318-675-6086. Fax: 1-318-675-6059.**

Diabetic mellitus is a serious and growing health problem worldwide. Increasing evidence from both clinical and experimental studies suggests the beneficial effect of fruit and vegetable supplements as a complementary therapeutic approach in diabetes. As a consequence, the demand for natural products has increased worldwide in an attempt to provide better treatment for diabetes. Compounds containin a thiol moiety are known to be good antioxidants. L-cysteine (LC) is one of the most well known biothiols, and it is utilized in the synthesis of various important signaling molecules including glutathione (GSH) and hydrogen sulfide (H_2S). Recent studies have shown that diabetic patients have reduced blood levels of LC and H_2S. Commonly eaten fruits and vegetables are rich sources of LC and H_2S. This review discusses the beneficial role of LC and H_2S rich fruits and vegetables in treatment of diabetes and its associated complications.

© 2013 American Chemical Society

Diabetes, Oxidative Stress, and Antioxidants

Diabetes has become an epidemic and remains a major public health issue worldwide. Intensive blood glucose control dramatically reduces the devastating complications that result from poorly controlled diabetes. However, for many patients, achieving tight glucose control is difficult with current regimens. This review discusses the scientific relevance for the regular consumption of certain fruits and vegetables as a novel adjuvant therapy to achieve better control of glycemia and improve the lives of the diabetic patient population.

The prevalence of oxidative stress in the pathogenesis of diabetes and its associated complications is well established (*1*, *2*), which raises the question whether antioxidants as therapeutics can prevent or delay the onset of the complications common to diabetes. Many different types of antioxidants, such as vitamin C, vitamin E, β-carotene, lipoic acids, and some metal based synthetic compounds (vanadium, selenium, chromium, and zinc) have been studied in cell culture models, animal models, and in diabetic patients. It has been reported that the therapeutic action of these compounds in the treatment of diabetes and its complications is mediated mostly via decreasing the rate of formation of reactive intermediates (*3*). Recently, there has been considerable interest in finding antioxidants from plant materials to replace the synthetic ones. Studies in the literature report that plants contain a large variety of antioxidant substances with fewer undesirable side effects (*4*, *5*). Phytochemicals found in fruits seem to enhance the nutritional value of these foods and they provide potent antioxidant power to combat oxidative stress and organ dysfunction (*6*). Biothiols or biologically derived thiols are the most important antioxidants that protect cells from the onset of oxidative stress (*7*).

Sources and Functions of Antioxidant Biothiols and L-Cysteine

L-cysteine (LC), an important semi-essential amino acid, plays a crucial role in controlling protein structure and stability (*8*). It is the most abundant biothiol and, in addition to its role in protein synthesis, cysteine is also utilized in the synthesis of coenzyme A, taurine, glutathione, and H_2S (*9*). A number of prospective observational studies have shown that hyperhomocysteinemia (HHcy), an elevated plasma homocysteine (Hcy) level, is an independent risk factor for cardiovascular disease (CVD) (*10*, *11*). Treatment with LC (*12*, *13*) or its derivative N-acetylcysteine (NAC) (*14*, *15*) or H_2S (*16*) has been found to be effective in lowering the plasma Hcy level and the risk of CVD.

Table I gives the LC concentration in different fruits and vegetables (Table I) (*17*). Red pepper has been shown to contain a high concentration of LC (349 nM/g wet weight). In addition, asparagus, spinach, green beans, and tomato are also reported to be cysteine rich vegetables. Among the commonly eaten fruits, orange, papaya, and strawberry have significantly high cysteine concentrations. Citrus juices, especially orange juice, have been suggested as good sources of cysteine by several health and nutrition groups since their intake is associated with improved lipid profiles and a reduced risk of cardiovascular disease (*18*, *19*)

Furthermore, orange juice is a rich source of flavonoids and vitamin C (*20*, *21*), which may scavenge ROS and suppress inflammatory processes. According to MayoClinic.com (MayoClinic.com: Dietary Fiber: Essential for A Healthy Diet), dietary fiber such as that found in oranges may slow down digestion so that blood sugar levels do not rise quickly after a meal. Due to its low glycemic index (less than 55), an orange slows the release of glucose into the blood stream. High cysteine content also makes the orange seem to be a good food for the prevention of diabetes and its complications.

Vegetables of the *Allium* family, which includes garlic, onions, leeks, shallots, etc. are known to be a rich source of sulfides (*22*). These vegetables contain S-alk(en)yl-L-cysteine sulfoxides within vacuoles of the plant cells. When the cells are disrupted by cutting or chewing, the vacuoles are broken, causing the enzymatic degradation of cysteine sulfoxides to sulfenic acids, which decomposes spontaneously to thiosulfinates. Thiosulfinates are unstable and decay into a complex mixture of compounds in which mono-, di-, and poly-sulfides predominate (*22*). Garlic and its relatives contain mainly S-allyl-L-cysteine sulfoxide, along with its decomposition products, such as diallyl sulfide, diallyl disulfide, diallyl trisulfide, and diallyl tetrasulfide. The predominant compound in onions and related *Alliums* are S-propyl-L-cysteine sulfoxide and S-prop-1-enyl-L-cysteine sulfoxide, and these yield the corresponding dipropyl sulfides, dipropyl disulfide, dipropyl trisulfide, dipropyl tetrasulfide, and diprop-1-enyl sulfides (*23*). Epidemiological studies suggest that consumption of *Allium* vegetables may protect against cancer in humans, although more intervention trials are necessary (*24*). Plants of the *Brassica* family (cabbage, Brussels sprouts, cauliflower, kale, etc.) contain S-methyl cysteine-L-sulfoxide, which upon degradation yields methyl sulfides (*25*). An increasing number of studies in the literature also report the efficacy of garlic in reducing blood glucose in various animal models of type 1 diabetes (*26*, *27*). In addition, the potential lipid lowering effect of garlic shows it to be a promising natural agent for the prevention of diabetic cardiac disorders such as atherosclerosis (*28*, *29*). The beneficial effect of garlic has been attributed to the presence of allicin and sulfur containing compounds (*30*). Using a high fructose fed (65%) diabetic male Sprague Dawley rat model, Padiya *et al.* (*31*) reported that administration of garlic homogenate at a dose of 250 mg/kg body weight/day for 8 weeks reduced hyperglycemia, insulin resistance, and oxidative stress, as well as boosting the serum H_2S level.

Soybeans provide an excellent source of dietary protein. Soybean protein contains ample amounts of all the essential amino acids plus several other macronutrients with a high nutritional value (*32*). Very recently Kim *et al.* (*33*) reported a successful strategy to increase the cysteine content of soybean seed through the overexpression of a key sulfur assimilatory enzyme, O-acetylserine sulfhydrylase (OASS). Clinical trials have reported the lipid-lowering effect of consuming soy products, and epidemiological studies have shown that dietary intake of soy products is associated with decreased risk of type 2 diabetes mellitus (T2DM) (*34*, *35*). Dietary supplementation with sesame flour protein also decreases the susceptibility to oxidative stress in hypercholesterolemic animals (*36*, *37*). The protein concentration of sesame seed is approximately 25%, with a range of 17-31% depending upon the source of the seed. Sesame protein is low in lysine content (3.1% protein), but it is rich in the sulfur containing amino acids methionine and cystine (6.1%), which are often the limiting amino acids in legumes (*38*). It has been reported that sesame oil exhibits a synergistic effect with anti-diabetic medication in patients with type 2 diabetes mellitus (*39*). Based upon these studies it can be speculated that to regulate diabetic complications, a suitable functional food could be made based upon the protein hydrolysate from oilseeds such as soybean, sesame, etc.

Oxidative stress plays a central role in the pathogenesis of diabetic complications. The presence of a free thiol group makes LC a potential antioxidant molecule. Any bioactive peptide containing LC as one of its amino acids has the potential to be an antioxidant therapeutic used in the prevention of oxidative stress and its associated organ dysfunction. The antioxidant property of LC is well known and thus it could be considered as an antidiabetic therapeutic also. Although there are ample antioxidants available, such as vitamin C, vitamin E, etc. at present there are no studies in the literature comparing the antioxidant activities of LC, vitamin C, or vitamin E.

Beneficial Role of H_2S and Its Precursor LC in Diabetes

Recent studies have shown that diabetic patients have reduced blood levels of hydrogen sulfide (H_2S) (*40*, *41*). H_2S is gaining acceptance as an important signaling molecule and has been shown to modulate a variety of biological effects that may mediate the protection of various organ dysfunctions (*42*, *43*). H_2S inhibits oxidative stress, promotes stimulation of K_{ATP} channels and relaxation and vasodilation in vascular smooth muscle cells, and relaxation of the human corpus cavernosum smooth muscle (*44–49*). *In vivo*, H_2S has been shown to inhibit leukocyte endothelial cell interactions and ischemia-reperfusion injury in liver and heart in animal studies (*46*). In mammalian systems, H_2S is produced from L-cysteine via the action of two enzymes, cystathionine-β-synthase (CBS) and cystathionine-γ-lyase (CSE) (*44*, *50*) (Figure 1). Both of these enzymes are dependent on pyridoxal-5′-phosphate. CSE is mainly expressed in the thoracic aorta, portal vein, ileum, heart, liver, kidney, and vascular smooth muscle, whereas CBS is highly expressed in the central and peripheral nervous systems (*44*, *50*, *51*). Recent studies have shown evidence for two other H_2S-producing enzymes, 3-mercaptopyruvate sulfurtransferase (3MST), and cysteine aminotransferase (CAT), which produce H_2S in the brain as well as in vascular endothelium (*45–47*). Human blood contains a significant amount of H_2S (10-100 μM) (*44*, *52*).

Table I. Cysteine Concentration (nM/g wet weight) in Common Vegetables and Fruits as Reported by Demirkol *et al.* (*17*)a

	Name	Cysteine (nM/g Wet Weight)
Vegetables	Asparagus	122±1
	Spinach	84±2
	Broccoli	ND
	Cauliflower	7±1
	Green squash	6±1
	Yellow squash	27±6
	Cucumber	11±3
	Carrot	ND
	Parsley	8±1
	Tomato	55±3
	Red pepper	349±18
	Green pepper	9±1
	Potato	ND
	Avocado	4±1
	Green beans	67±11
Fruits	Orange	41±2
	Lemon	6±0
	Grapefruit	15±2
	Mango	10±0
	Papaya	58±5
	Banana	7±0
	Strawberry	59±5

[a] The values are given as mean±SD (n=3). ND: Not detectable.

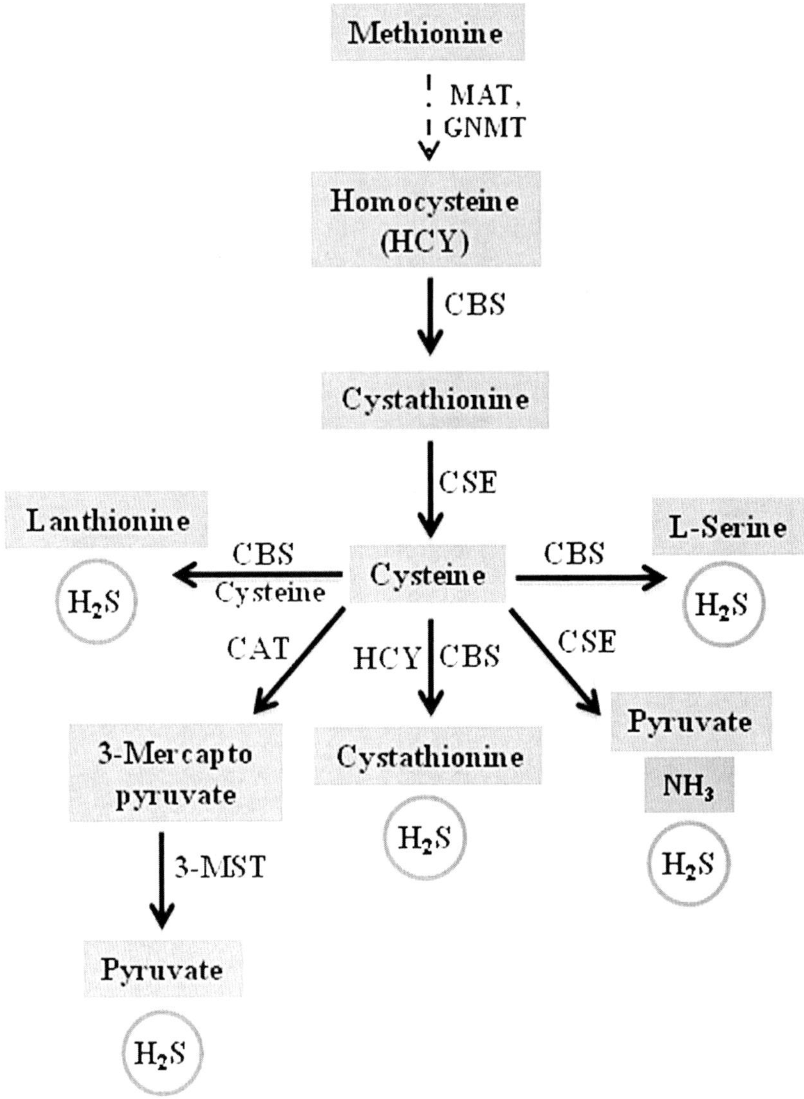

Figure 1. Schematic representation of the formation of H_2S from cysteine.

Diabetic patients experience altered cysteine homeostasis, which causes lower blood levels of LC (*53*, *54*). A number of previous studies have shown that dietary supplementation with NAC (N-acetyl cysteine) or whey protein and α-lactoalbumin (cysteine rich proteins) lowers the oxidative stress, and insulin resistance induced by sucrose or fructose in rats and streptozotocin-treated diabetic mice (*55–59*). Oral supplementation with LC lowered glycemia, oxidative stress and vascular inflammation markers in ZDF rats, an animal model of type 2 diabetes (*60*). Studies with type 2 diabetic patients and normal subjects have shown that LC supplementation lowered oxidative stress markers in the diabetic population compared to those in the normal population (*53*, *54*). Sekhar *et al.* (*61*) showed that dietary supplementation with LC lowered the fasting plasma glucose levels in elderly human subjects. These studies led to a novel hypothesis that reduced LC or H_2S levels may play a direct role in modulating glucose metabolism in diabetes. However, studies on the molecular mechanism by which LC or H_2S increases glucose utilization and lowers glycemia is needed.

PI3K (phosphoinositide-3-kinase) and PTEN (phosphatase and tensin homologue) play central roles in the insulin signaling cascade and glucose metabolism (*62*). The activation of Class I PI3K causes phosphorylation of PIP2 at position 3 of its inositol head group, which leads to the formation of PIP3 (phosphatidylinositol-3,4,5-triphosphate), but PTEN activation causes the degradation of PIP3 (*63*). This means that the cellular PIP3 concentration is regulated by the PI3K/PTEN equilibrium (Figure 2). Studies in the literature have shown that inhibition of PTEN expression using PTEN antisense oligonucleotides normalized blood glucose levels in ob/ob (obese) mice (*64*) and that overexpression of PTEN resulted in inhibition of insulin-induced PIP3 production and glucose uptake in 3T3L1 adipocytes (*65*). Cells that lack PTEN showed elevated PIP3 levels and activated AKT-dependent signaling (*65*). Comparative signal silencing studies using antisense AKT2 (serine/threonine protein kinase B) and antisense PKCζ (protein kinase Cζ) demonstrated that PKCζ is significantly more active in PIP3 mediated GLUT4 (glucose transporter 4) activation and glucose uptake, as well as in glucose utilization, using a 3T3L1 adipocyte cell model (*66*). All of these findings suggested that PIP3 plays a major role in the insulin signaling pathway (*63*, *67*) (Figure 2).

Using ZDF rats, a model of type 2 diabetes, Jain *et al.* (*60*) reported that LC supplementation caused activation of PI3K and inhibition of NF-κB in the liver and reduction in plasma glucose levels. From further studies using an adipocyte cell model, Manna and Jain (*66*) reported that LC supplementation caused PI3K activation, PTEN inhibition and an increase in intracellular PIP3 (phosphatidylinositol-3,4,5 trisphosphate) levels and glucose utilization in cells exposed to high glucose (HG). The effect of LC on PIP3 and glucose utilization was prevented by PAG (propargylglycine), an inhibitor of cystathionine-γ-lyase (CSE) that catalyzes H_2S formation from LC. Treatment with LC, H_2S, or PIP3 increased the phosphorylation of IRS1, AKT, and PKCζ/λ, as well as GLUT4 activation and glucose utilization in HG-treated cells. Comparative signal silencing studies using antisense AKT2 and antisense PKCζ demonstrated that PKCζ is significantly more active in PIP3 mediated GLUT4 activation and glucose utilization adipocytes exposed to HG. These studies provide evidence for a novel molecular mechanism by which LC or H_2S can increase PIP3 and upregulate the metabolic actions of insulin, thus leading to improved glucose metabolism (Figure 3).

This review provides the molecular mechanism and the scientific relevance supporting the regular consumption of certain fruits and vegetables as a novel adjuvant therapy to improve glucose metabolism and prevent CVD in the diabetic patient population.

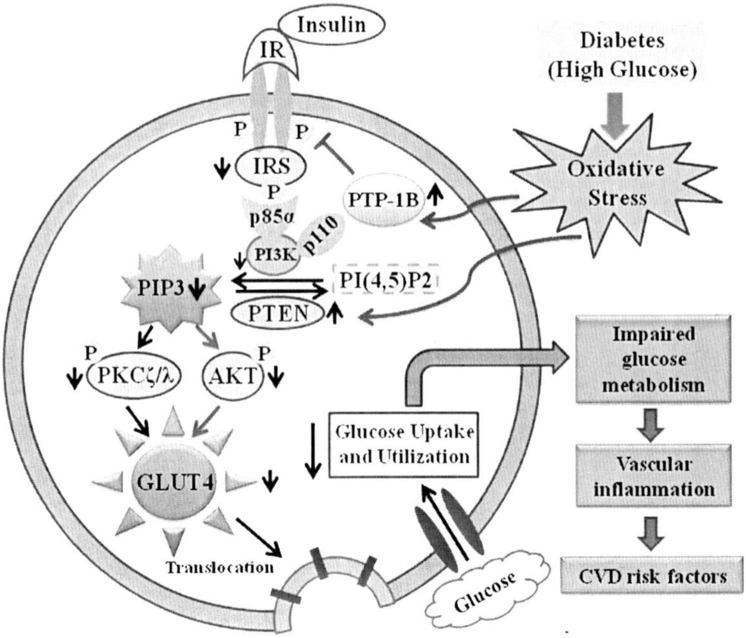

Figure 2. Schematic representation of the signaling pathways involved in insulin stimulated glucose utilization and the effect of a diabetic episode on it.

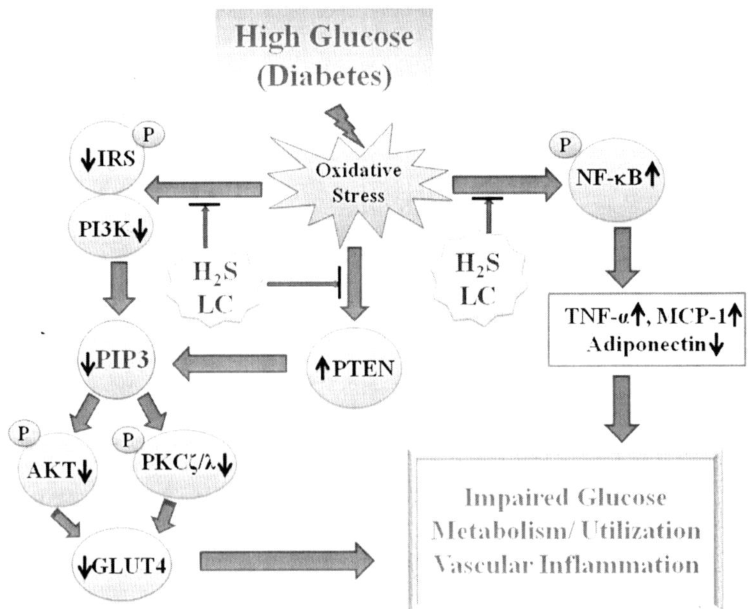

Figure 3. Schematic diagram of the proposed mechanism by which H_2S or LC affects glucose metabolism via inhibition of PTEN/NF-κB and upregulation of PI3K/PIP3/AKT/PKCζ/λ signaling cascades.

Acknowledgments

The authors are supported by grants from NIDDK and the Office of Dietary Supplements of the National Institutes of Health RO1 DK072433 and the Malcolm Feist Endowed Chair in Diabetes. This study is also funded by a fellowship from the Malcolm Feist Cardiovascular Research Endowment, LSU Health Sciences Center, Shreveport. The authors thank Ms Georgia Morgan for excellent editing of this manuscript.

References

1. West, I. C. *Diabet. Med.* **2000**, *17*, 171–180.
2. Brownlee, M. *Nature* **2001**, *414*, 813–820.
3. Rahimi, R.; Nikfar, S.; Larijani, B.; Abdollahi, M. *Biomed. Pharmacother.* **2005**, *59*, 365–373.
4. Krishnaiah, D.; Sarbatly, R.; Nithyanandam, R. *Food Bioprod. Process.* **2011**, *89*, 217–233.
5. Scartezzini, P; Speroni, E. Review on some plants of Indian traditional medicine with antioxidant activity. *J. Ethnopharmacol.* **2000**, *71*, 23–43.
6. Sun, J.; Chu, Y. F.; Wu, X.; Liu, R. H. *J. Agric. Food. Chem.* **2002**, *50*, 7449–7454.
7. Wlodek, L. *Pol. J. Pharmacol.* **2002**, *54*, 215–223.
8. Brosnan, J. T.; Brosnan, M. E. *J. Nutr.* **2006**, *136*, 1636S–1640S.
9. Stipanuk, M. H.; Dominy, J. E.; Lee, J. I.; Coloso, R. M. *J. Nutr.* **2006**, *136*, 1652S–1692S.
10. Vollset, S. E.; Refsum, H.; Tverdal, A.; Nygard, O.; Nordrehaug, J. E.; Tell, G. S.; Ueland, P. M. *Am. J. Clin. Nutr.* **2001**, *74*, 130–136.
11. Clarke, R.; Collins, R.; Lewington, S.; et al. *JAMA, J. Am. Med. Assoc.* **2002**, *288*, 2015–2022.
12. Okawa, H.; Morita, T.; Sugiyama, K. *Biosci. Biotechnol. Biochem.* **2007**, *71*, 91–97.
13. Verhoef, P.; Steenge, G. R.; Boelsma, E.; van Vliet, T.; Olthof, M. R.; Katan, M. B. *Am. J. Clin. Nutr.* **2004**, *80*, 674–679.
14. Ventura, P.; Panini, R.; Pasini, M. C.; Scarpetta, G.; Salvioli, G. *Pharmacol. Res.* **1999**, *40*, 345–350.
15. Wiklund, O.; Fager, G.; Andersson, A.; Lundstam, U.; Masson, P.; Hultberg, B. *Atherosclerosis* **1996**, *119*, 99–106.
16. Sen, U.; Givvimani, S.; Abe, O. A.; Lederer, E. D.; Tyagi, S. C. *Am. J. Physiol. Cell. Physiol.* **2011**, C155–163.
17. Demirkol, O.; Adams, C.; Ercal, N. *J. Agric. Food. Chem.* **2004**, *52*, 8151–8154.
18. Kurowska, E. M.; Spence, J. D.; Jordan, J.; Wetmore, S.; Freeman, D. J.; Piche, L. A.; Serratore, P. *Am. J. Clin. Nutr.* **2000**, *72*, 1095–1100.
19. Liu, S.; Manson, J. E.; Lee, I. M.; Cole, S,R.; Hennekens, C. H.; Willett, W. C.; Buring, J. E. *Am. J. Clin. Nutr.* **2000**, *72*, 922–928.
20. Squires, S. R.; Hanna, J. G. *J. Agric. Food. Chem.* **1979**, *27*, 639–641.
21. Nelson, E. W.; Streiff, R. R.; Cerda, J. J. *Am. J. Clin. Nutr.* **1975**, 1014–1019.

22. Lawson, L. D. In *Phytomedicines of Europe. Chemistry and Biological Activity*; Lawson, L. D., Bauer, R., Eds.; ACS Symposium Series 691; American Chemical Society: Washington, DC, 1998; pp 176–209.
23. Block, E. In *Garlic and Other Alliums: The Lore and the Science*; RSC Publishing: Cambridge, U.K., 2009.
24. Zhou, Y.; Zhuang, W.; Hu, W.; Liu, G. J.; Wu, T. X.; Wu, X. T. *Gastroenterology* **2011**, *141*, 80–89.
25. Smith, R. H. *Vet. Sci. Comm.* **1978**, *2*, 47–61.
26. Farva, D.; Goji, L. A.; Joseph, P. K.; Augusti, K. T. *Ind. J. Biochem. Biophys.* **1986**, *23*, 24–27.
27. Ohaeri, O. C. *Bio. Sci. Rep.* **2001**, *2*, 19–24.
28. Sobenin, I. A.; Andrianova, I. V.; Demidova, O. N.; Gorchakova, T.; Orekhov, A. N. *J. Atheroscler. Thromb.* **2008**, *15*, 334–338.
29. Sener, G.; Sakarcan, A.; Yegen, B. C. *Mol. Nutr. Food Res.* **2007**, *51*, 1345–1352.
30. Mathew, P. T.; Augusti, K. T. *Ind. J. Biochem. Biophys.* **1973**, *10*, 209–212.
31. Padiya, R.; Khatua, T. N.; Bagul, P. K.; Kuncha, M.; Banerjee, S. K. *Nutr. Metab. (Lond).* **2011**, *8*, 53.
32. Velasquez, M. T.; Bhathena, S. J. *Int. J. Med. Sci.* **2007**, *4*, 72–82.
33. Kim, W. S.; Chronis, D.; Juergens, M.; et al. *Planta* **2012**, *235*, 13–23.
34. Yang, B.; Chen, Y.; Xu, T.; Yu, Y.; Huang, T.; Hu, X.; Li, D. *Asia Pac. J. Clin. Nutr.* **2011**, *20*, 593–602.
35. Anderson, J. W.; Johnstone, B. M.; Cook-Newell, M. E. *N. Engl. J. Med.* **1995**, *333*, 276–282.
36. Kang, M. H.; Kawai, Y.; Naito, M.; Osawa, T. *J. Nutr.* **1999**, *129*, 1885–1890.
37. Korou, L. M.; Agrogiannis, G.; Pantopoulou, A.; et al. *Lipids Health Dis.* **2010**, *9*, 23.
38. Hwang, L. S. In *Bailey's Industrial Oil and Fat Products*, 6th ed.; Shahidi, F., Ed.; John Wiley & Sons, Inc.: Hoboken, NJ, 2005; Vol. 6, pp537–576.
39. Sankar, D.; Ali, A.; Sambandam, G.; Rao, R. *Clin. Nutr.* **2011**, *30*, 351–358.
40. Jain, S. K.; Bull, R.; Rains, J. L.; et al. *Antioxid. Redox Signal* **2010**, *12*, 1333–1337.
41. Whiteman, M.; Gooding, K. M.; Whatmore, J. L.; et al. *Diabetologia* **2010**, *53*, 1722–1726.
42. Szabó, C. *Nat. Rev. Drug. Discov.* **2007**, *6*, 917–35.
43. Whiteman, M.; Le Trionnaire, S.; Chopra, M.; Fox, B.; Whatmore, J. *Clin. Sci. (Lond)* **2011**, *121*, 459–88.
44. Yang, G.; Wu, L.; Jiang, B.; et al. *Science* **2008**, *322*, 587–590.
45. Ishigami, M.; Hiraki, K.; Umemura, K.; Ogasawara, Y.; Ishii, K.; Kimura, H. *Antioxid. Redox Signal* **2009**, *21*, 205–214.
46. Shibuya, N.; Tanaka, M.; Yoshida, M.; et al. *Antioxid. Redox Signal* **2009**, *11*, 703–714.
47. Shibuya, N.; Mikami, Y.; Kimura, Y.; Nagahara, N.; Kimura, H. *J. Biochem.* **2009**, *146*, 623–626.
48. d'Emmanuele di Villa Bianca, R.; Sorrentino, R.; Maffia, P.; et al. *Proc. Natl. Acad. Sci. U.S.A.* **2009**, *106*, 4513–4518.
49. Kimura, Y.; Goto, Y; Kimura, H. *Antioxid. Redox Signal* **2010**, *12*, 1–13.

50. Jha, S.; Calvert, J. W.; Durankski, M. R.; Ramachandran, A.; Lefer, D. J. *Am. J. Physiol. Heart. Circ. Physiol.* **2008**, *295*, H801–H806.
51. Singh, S.; Padovani, D.; Leslie, R. A.; Chiku, T.; Banerjee, R. *J. Biol. Chem.* **2009**, *284*, 22457–22466.
52. Cai, W. J.; Wang, M. J.; Moore, P. K.; Jin, H. M.; Yao, T.; Zhu, Y. C. *Cardiovasc. Res.* **2007**, *76*, 29–34.
53. Darmaun, D.; Smith, S. D.; Sweeten, S.; Hartman, B. K.; Welch, C.; Mauras, N. *Pediatr. Diabetes* **2008**, *9*, 577–582.
54. Sekhar, R. V.; McKay, S. V.; Patel, S. G.; Guthikonda, A. P.; Reddy, V. T.; Balasubramanyam, A.; Jahoor, F. *Diabetes Care* **2011**, *34*, 162–167.
55. Blouet, C.; Mariotti, F.; Mikogami, T.; Tome, D.; Huneau, J. F. *J. Nutr. Biochem.* **2007**, *18*, 519–524.
56. Blouet, C. B.; Mariotti, F.; Azzout-Marniche, D.; et al. *Free Radical Biol. Med.* **2007**, *42*, 1089–1097.
57. Diniz, Y. S.; Rocha, K. K.; Souza, G. A.; Galhardi, C. M.; Ebaid, G. M.; Rodrigues, H. G.; et al. *Eur. J. Pharmacol.* **2006**, *543*, 151–157.
58. Haber, C. A.; Lam, T. K.; Yu, Z.; Gupta, N.; Goh, T.; Bogdanovic, E.; et al. *Am. J. Physiol. Endocrinol. Metab.* **2003**, *285*, E744–753.
59. Song, D.; Hutchings, S.; Pang, C. C. Y. *Eur. J. Pharmacol.* **2005**, *508*, 205–210.
60. Jain, S. K.; Velusamy, T.; Croad, J. L.; Rains, J. L.; Bull, R. *Free Radical Biol. Med.* **2009**, *46*, 1633–1638.
61. Sekhar, R. V.; Patel, S. G.; Guthikonda, A. P.; Reid, M.; Balasubramanyam, A.; Taffet, G. E.; Jahoor, F. *Am. J. Clin. Nutr.* **2011**, *94*, 847–853.
62. Taniguchi, C. M.; Tran, T. T.; Kondo, T.; et al. *Proc. Natl. Acad. Sci. U.S.A.* **2006**, *103*, 12093–12097.
63. Pendaries, C.; Tronchère, H.; Plantavid, M.; Payrastre, B. *FEBS Lett.* **2003**, *546*, 25–31.
64. Butler, M.; McKay, R. A.; Popoff, I. J. *Diabetes* **2002**, *51*, 1028–1034.
65. Ono, H.; Katagiri, H.; Funaki, M.; et al. *Mol. Endocrinol.* **2001**, *15*, 1411–1422.
66. Manna, P.; Jain, S. K. *J. Biol. Chem.* **2011**, *286*, 39848–39859.
67. Hinchliffe, K. A. *Curr. Biol.* **2001**, *11*, R371–R372.

Chapter 10

Issues Surrounding the Anti-Inflammatory Actions of the Citrus Polymethoxylated Flavones

John A. Manthey*,[1] and Thaïs B. César[2]

[1]U.S. Horticultural Research Laboratory, USDA-ARS, Ft. Pierce, Florida 34945
[2]Department of Food and Nutrition, Faculty of Pharmaceutical Sciences-UNESP, Araraquara, SP, Brazil
***E-mail: John.Manthey@ars.usda.gov.**

The polymethoxylated flavones in citrus have been evaluated for their *in vivo* anti-inflammatory actions in several animal assays. Strong anti-inflammatory effects were observed following administration of 3,5,6,7,8,3′,4′-heptamethoxyflavone (HMF) dissolved in vegetable oil by intraperitoneal (i.p.) injection, but neglibible action was detected following doses of HMF suspended as fine particulates in vegetable oil. This contradiction is shown to be a reflection of differences in the bioavailabilities of HMF following different routes of administration. Evidence of contrasting bioavailabilities among different PMFs is shown through comparisons of the pharmacokinetics of nobiletin and tangeretin administered in rats.

Introduction

The first major work with flavonoids relative to inflammation was the series of papers by Szent-Györgyi and co-workers, where they reported on the capillary sparing properties of flavonoids in their efforts to establish vitamin-like properties of these compounds (*1–3*). In their first papers they reported that flavonoids of

© 2013 American Chemical Society

red peppers and lemons increased the resisitance of capillaries to bleeding in patients with certain pathological conditions. Additional supportive findings were subsequently made with guinea pigs, and the term vitamin P, or the permeability vitamin was suggested. Yet, there was great difficulty in consistently observing a vitamin deficiency strictly attributable to flavonoid deficiency, and in 1950 the vitamin P designation was dropped (*4*). But, what was not dropped was the interest in these compounds, particularly in their possible anti-inflammatory actions. In fact, flavonoids have now been extensively studied for their effects in nearly every area of human biology (*5*, *6*).

From these earliest studies of Szent-Györgyi to the present, there has been a strong interest in the citrus flavonoids, partly because of the marvelous diversity of structures for these compounds in citrus. Amongst these are the highly methoxylated flavone aglycones, which are termed the polymethoxylated flavones (PMFs), and include such compounds as tangeretin, nobiletin, and heptamethoxyflavone (Figure 1). There are numerous others, and a particularly complete description of these compounds has been reported (*7*). There are also high concentrations of flavanone glycosides, such as hesperidin, naringin and eriocitrin, in oranges, grapefruit and lemons, respectively (*8*). The description of hesperidin is of particularly early origin, being first reported by Lebreton in 1828 (*9*). In addition to these compounds, there are also flavone-*O*-glycosides, i.e. diosmin, and flavone-*C*-glycosides, i.e. 6,8-di-*C*-glucosylapigenin (Figure 1). The diversity of structures extends to many other compounds, reported to a large extent by Matsubara and co-workers ((*10*) and references therein).

In focusing on the PMFs, we find that one of the first main studies of their possible anti-inflammatory effects is an investigation by Freedman and Merritt 1963 (*11*), where they reported potent inhibition of the inflammation responses in the Arthus reaction inflammation model. These researchers fractionated a citrus bioflavonoid complex derived from combined orange and grapefruit peel into 3 separate portions, and found that 65 percent of the total anti-inflammatory activity of the whole complex was associated with a fraction that contained hesperidin, naringin and nobiletin (NOB), plus an undetermined pentamethoxyflavone (sinensetin?), and an unknown compound labeled RS-1. These individual compounds were administered in oral doses to guinea pigs subjected to the Arthus reaction. Results showed that hesperidin and naringin were inactive, while NOB, the unknown RS-1, and the undetermined pentamethoxyflavone exhibited ED_{25} values lower than the positive control, hydrocortisone phosphate. Unfortunately, no further information about the identities of the undetermined pentamethoxyflavone and of the unknown compound RS-1 appeared in the literature. Yet, evidence reported elsewhere continued to substantiate the anti-inflammatory properties of other orally administered citrus flavonoids, particularly the flavanone and flavone glycosides, hesperidin and diosmin, the two main compounds in the commercial product Daflon 500 mg. A very solid body of information now exists about the biological actions of these compounds in the inflammation cascade (*12–14*), and the commercial product, Daflon 500 mg has been shown effective in a variety of clinical settings (*15–17*).

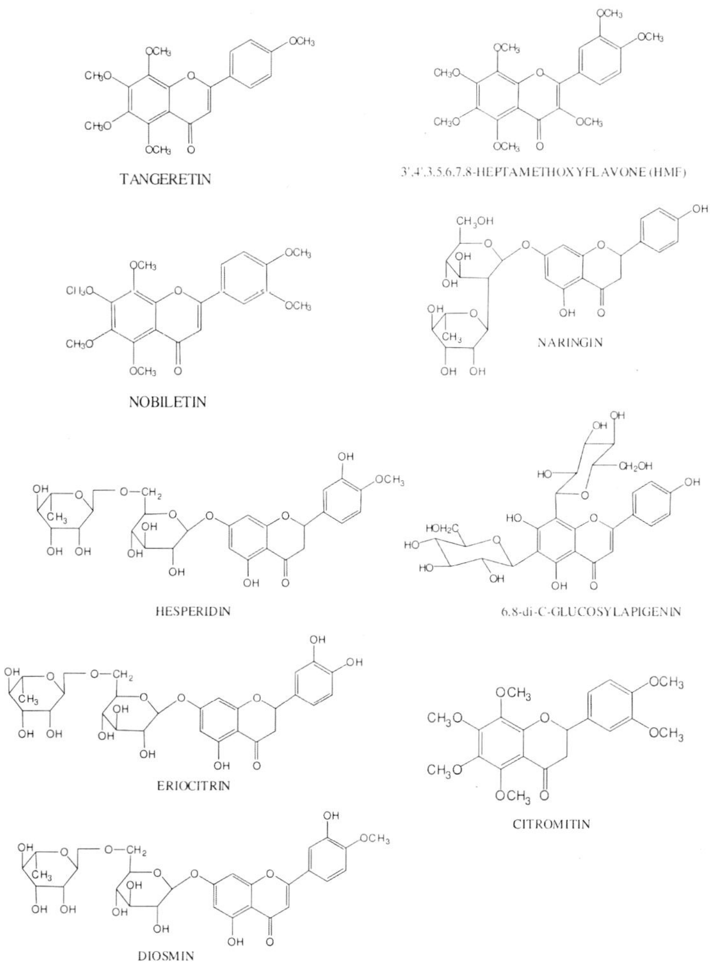

Figure 1. Structures of select flavonoids in citrus.

PMF in Vitro Study

This past decade has shown a renewed interest in the biological actions of the PMFs, and our participation in this work initially involved a collaborative study which looked at the influences of these compounds on *in vitro* bacterial lipopolysaccharide (LPS) activation of human monocytes (*18*). This LPS activation of monocytes into macrophages mimics a central step in the inflammation cascade, where activation of monocytes triggers the production of extremely pro-inflammatory cytokine proteins, some of which include the tumor necrosis factor-α (TNF-α), macrophage inflammatory protein (MIP-1α) and sets of interleukins. In this *in vitro* study, it was discovered that the citrus PMFs were able to block the LPS-induced production of TNF-α, MIP-1α and of interleukin-10, while the LPS-induced production of other interleukins, including IL-1β, 6 and 8, was unaffected. Evidence supported the conclusion that the PMFs influence cytokine production at the level of gene transcription, rather than resulting from a general cytotoxicity of these compounds. Much of the study focused on the compound, 3,5,6,7,8,3′,4′-heptamethoxyflavone (HMF), which exhibited one of the the lowest IC_{50} values (5 μM) towards the inhibition of LPS-induced production of TNF-α, and which was also shown to be a potent inhibitor of human phosphodiesterase-4 activity, leading to substantial increases in cAMP levels in the human monocytes (*18*). Inhibition of phosphodiesterase activity, along with increased cAMP levels, is a known link with decreased cytokine production (*19*). Inhibition of phosphodiesterase activities by methoxylated flavones was earlier observed by Nikaido et al. (*20*), and Petkov et al., (*21*), and the findings with HMF were in agreement with these earlier reports. This suggested a reasonable mode of action by which the PMFs modulate inflammation *in vitro*, but it remained to be determined whether these compounds could also exhibit similar anti-inflammatory activities *in vivo*.

PMF in Vivo Studies

This question about the *in vivo* actions of HMF was addressed in two common inflammation models, the rat carrageenan/paw edema and the mouse LPS-challenge assays (*22*). This latter assay very closely mimics the *in vitro* LPS-challenged human monocyte assay discussed in the previous section. In this assay, bacterial LPS injected into mice elicits a massive production of TNF-α in the blood, an event typically fatal to the mice. In the first experiment, HMF dissolved in vegetable oil (VO) and administered by intraperitoneal (i.p.) injection into mice inhibited the LPS-induced TNF-α production by 30% at 33 mg/kg and 45% at 100 mg/kg. Dexmethasone, the positive control, produced a 66% inhibition at 0.5 mg/kg (Table I). Also shown in Table I are the HMF concentrations in the blood sera of the mice at 2.5 hours after HMF injections. With 33 mg/kg dissolved HMF the levels averaged 8.2 ppm and for the dose of 100 mg/kg the levels averaged 14.2 ppm. These serum levels of HMF suggest effective transfer of HMF to the blood resulting from these i.p. injections of dissolved HMF.

In the next experiment, HMF was administered as finely suspended particulates in VO, rather than dissolved in VO. The suspended particulates were given by i.p. injection at 100 mg/kg or orally by gavage at 50 to 200 mg/kg. The results showed that HMF given as a suspension under any of these conditions failed to inhibit TNF-α production in the LPS-challenged mice. There were no decreases in the LPS-induced TNF-α levels following the oral doses; in fact, the average values (2203-2399 pg/mL) resulting after these oral doses were slightly higher than for the "VO vehicle plus LPS" control treatment group (1920±678 pg/mL) (Table II). Injections (i.p.) of HMF suspended in VO at 100 mg/kg also produced negligible responses. Yet, the positive control prednisolone (10mg/kg (i.p.)), produced a powerful anti-inflammatory response.

Table I. Serum TNF-α Levels (pg/mL) Following LPS Challenge in Mice with HMF/VO Solution (ip) Administration. (Reproduced from *J. Agric. Food Chem.* 2008, *56*, 9399-9403)

Treatment	*Serum TNF-α Levels (pg/mL)*	*% Inhibition*	*[HMF]$_{serum}$*
VO vehicle	6167±182	0	
Dexmethasone (0.5 mg/kg)	2112±50	66[a]/42[b]	
HMF (33 mg/kg)	4313±144	30%[a]	8.2±1.5 ppm
HMF (100 mg/kg)	3411±86	45%[a]	14.2±4.7 ppm

[a] Relative to VO vehicle. [b] Relative to 50% PEG 400 vehicle.

Table II. Serum TNF-α Levels (pg/mL) Following LPS Challenge in Mice with HMF/VO Suspension (p.o.) and (i.p.) Administration. (Reproduced from *J. Agric. Food Chem.* 2008, *56*, 9399-9403)

Group	*Treatment*	*Serum TNF-α Levels (pg/mL)*	*p vs vehicle*	*[HMF]$_{serum}$*
1	VO vehicle minus LPS	70.3±35		
2	VO vehicle plus LPS	1920±678		
3	prednisolone (10 mg/kg (op) + LPS	617±143	0.001	
4	HMF (50 mg/kg (op)) + LPS	2399±843	0.232	
5	HMF (100 mg/kg (op)) + LPS	2286±653	0.290	0.035 ppm
6	HMF (200 mg/kg (op)) + LPS	2203±793	0.456	0.048 ppm
7	VO vehicle (ip) + LPS	2079±660		
8	Prednisolone (10 mg/kg (ip)) + LPS	656±177	0.0004	
9	HMF (100 mg/kg (ip) + LPS	1850±762	0.53	0.52 ppm

These results with HMF were not what was expected in light of the earlier findings of Freedman and Merritt (1963). Yet, one clue for this might be the extremely low levels of HMF in the blood system following the treatments with HMF suspended in VO. The 100 and 200 mg/kg oral doses reached only 0.035 and 0.048 ppm HMF blood serum levels, respectively, and even the i.p. dose of 100 mg/kg produced HMF levels of only 0.52 ppm (Table II). With such low serum levels of HMF it is not surprising that no inhibitory activity was observed.

The rat carrageenan/paw edema assay with HMF was also run, and similarly no inhibitory activity occurred resulting from oral doses of particulate HMF in VO, even up to 100 mg/kg (Table III). The oral dose of VO (minus HMF) produced an average weight difference of 0.94 g, while the oral doses of suspended HMF produced slight average weight increases of 1.07 and 1.11 g for the 50 mg/kg and 100 mg/kg doses, respectively. In contrast, in animals dosed with HMF *dissolved* in VO and given by i.p. injection, a significant amount of inhibition occurred, 56% at 100 mg/kg. This is in comparison to the 32% inhibition with the indomethacin positive control at 4 mg/kg by subcutaneous injection.

Table III. Right vs Left Paw Weight Differences (g) Following Carrageenan Injection in Rats with HMF/VO Suspension (p.o.) and (i.p.) Solution Injection. (Reproduced from *J. Agric. Food Chem.* 2008, *56*, 9399-9403)

Group	*Mean±SE*	*% Inhibition*
1. Water (po)	1.35±0.07	0%
2. VO (po)	0.94±0.06	30%
3. HMF (suspended in VO) 50 mg/kg (po)	1.11±0.07	18%
4. HMF (suspended in VO) 100 mg/kg (po)	1.07±0.07	21%
5. VO (ip)	1.14±0.10	16%
6. HMF (dissolved in VO) 100 gm/kg (ip)	0.60±0.06	56%
7. Indomethacin 4mg/kg (sc)	0.92±0.06	32%

As a furtherance of our study of HMF, we also ran the Arthus reaction, the same assay run by Freedman and Merritt (*11*), which was designed as an immune complex-driven inflammation model initially run in guinea pigs as a test for antirheumatoid drugs (*23–25*). The inflammation response in this older design is the swelling, or edema in the ankles where the treatments are given. In our experiments, the HMF, VO, or hydrocortisone (HC) treatments were given by either i.p. injection or by gavage 1 hour preceding initiation of the reaction. Inflammation was measured hourly by rear ankle swelling using digital calipers, and was measured over 6 hours.

The results in Figure 2 show the measurements of the ankle swelling resulting after oral doses of VO, 50 mg/kg HMF, 160 mg/kg HMF, and hydrocortisone (the positive control), where the HMF was dissolved in VO. The time courses in Figure 2 contrast the amounts of swelling occurring after oral doses of VO (minus HMF), and of oral doses of hydrocortisone. Also shown is the time course of ankle swelling resulting after an oral dose of 50 mg HMF/kg, which produced no inhibition of ankle swelling; but in fact caused slight increases in swelling. However, at the higher dose of 160 mg/kg there was a significant amount of inhibition of ankle swelling.

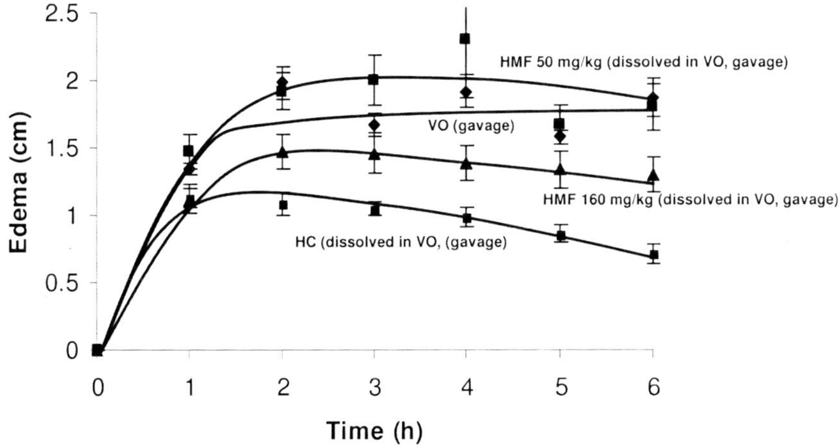

Figure 2. Ankle swelling time courses in guinea pigs following oral doses of VO (minus HMF), hydrocortisone (HC), and HMF (50 and 160 mg/kg) dissolved in VO.

The resulting inhibition of ankle swelling following doses of HMF administered by i.p. injection are shown in Figure 3. The dose of 200 mg/kg (dissolved in VO) was nearly as effective in inhibiting ankle swelling as the hydrocortisone positive control during first 3-4 hours. These findings with the Arthus reaction assay are consistent with the previous studies using a variety of other inflammation models as discussed above. Excellent anti-inflammatory activity of HMF occurs following doses of HMF dissolved in VO, but doses of HMF suspended as undissolved particulates in VO seem not to influence inflammation. Measurements of HMF blood serum levels indicate that this lack of effect on inflammation may likely be due to the low bioavailability of HMF given in this manner.

Pertinent to this latter observation are recent findings on the the bioavailabilities of other PMFs, mainly nobiletin (NOB) and tangeretin (TAN) (*26*). Figure 4 shows the time courses of the serum levels in rats of NOB following doses of dissolved or suspended NOB in VO. For an oral dose of 50 mg/kg NOB dissolved in VO, initial serum levels averaged 9.3 μg/mL serum. In contrast, the average maximum NOB concentration following oral doses of NOB suspended in VO occurred at 2 h and reached levels of nearly 3 μg/mL. Even this lower serum level (3 μg/mL) was far higher than the levels measured in mouse sera following oral doses of HMF (>0.048 μg/mL) (Table II). In a study of TAN, oral doses of 50 mg/kg TAN dissolved in VO produced average rat blood serum levels only as high as 0.43 μg/ml (Figure 5), far less than the serum level resulting after equivalent doses of NOB (Figure 4). These observations are summarized in Table IV. Admittedly, the values listed here are a compilation of measurements made using different dosing regimes and animal species, but the trends in Table IV suggest that amongst other factors, differences in oral availabilities may play important roles in influencing activity levels among the different PMFs, and if this is true, then this may explain (partly at least) the discrepancies between the findings that have been made with HMF and those of Freedman and Merritt (*11*) with NOB.

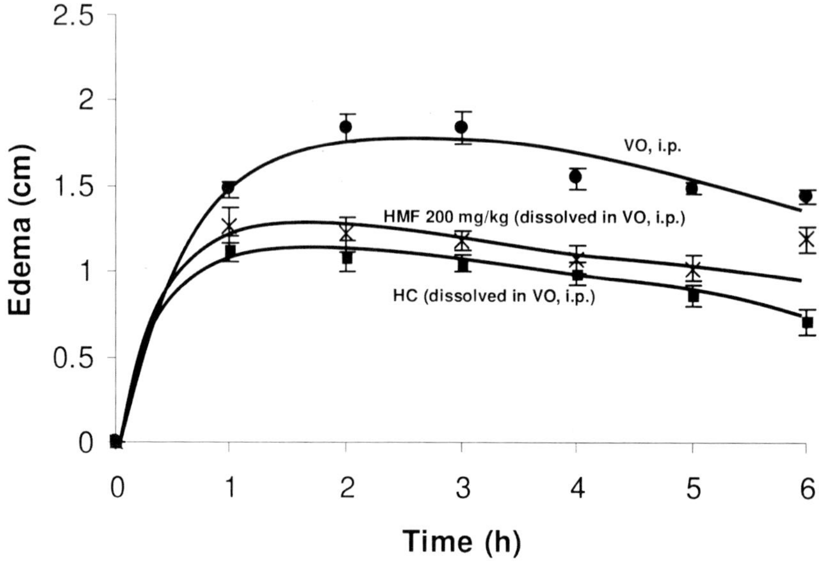

Figure 3. Ankle swelling time courses in guinea pigs following doses by i.p. injection of VO (minus HMF), hydrocortisone (HC), and HMF (200 mg/kg) dissolved in VO.

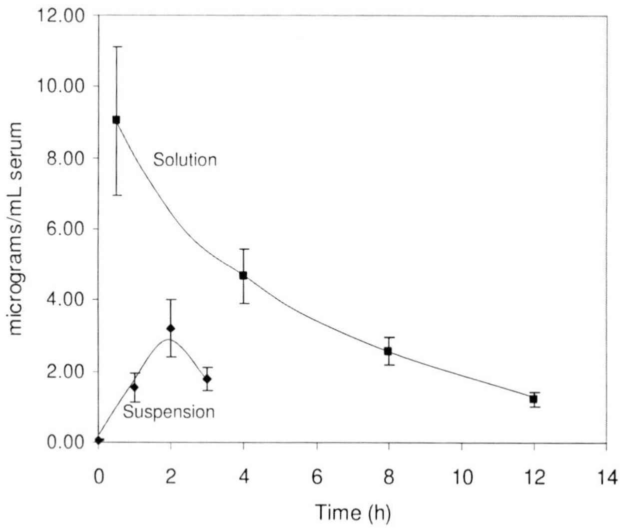

Figure 4. Pharmacokinetics of NOB administered at a dose of 50 mg kg^{-1} body weight to rats by gavage as a dissolved solution or as a suspension in corn oil. Reproduced from J. Agric. Food Chem. ***2011****, 59, 145-151.*

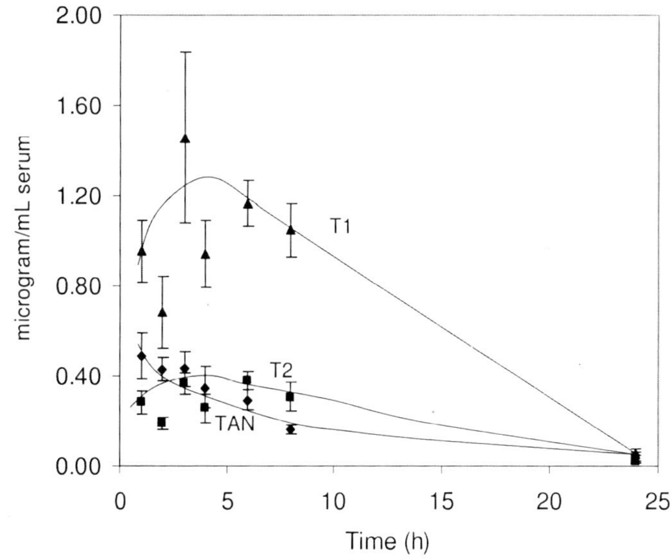

Figure 5. Pharmacokinetics of TAN and metabolites T1 and T2 in rat serum following administration by gavage at a dose of 50 mg kg^{-1} body weight. Reproduced from J. Agric. Food Chem. ***2011****, 59, 145-151.*

Table IV. Compilation of PMF Serum Levels Observed in Different Experiments, Different Animals, Different Modes, and Formulations

PMF	*mode*	*time*	*dose*	*animal*	*formulation*	*$[PMF]_{serum}$ (ppm)*
HMF	(i.p.)	2.5 h	33 mg.kg	mouse	dissolved	8.2
HMF	(i.p.)	2.5 h	100 mg/kg	mouse	dissolved	14.2
HMF	p.o.	2.5 h	100 mg/kg	mouse	particulate	0.035
HMF	p.o.	2.5 h	200 mg/kg	mouse	particulate	0.48
HMF	i.p.	2.5 h	100 mg/kg	mouse	particulate	0.52
TAN	p.o.	6 h	50 mg/kg	rat	dissolved	~0.25
TAN	p.o.	.5 h	50 mg/kg	rat	dissolved	~0.49
NOB	p.o.	0.5 h	50 mg/kg	rat	dissolved	~9.3
NOB	p.o.	6 h	50 mg/kg	rat	dissolved	~3.8

Acknowledgments

Mention of a trademark or proprietary product is for identification only and does not imply a guarantee or warranty of the product by the U.S. Department of Agriculture. The U.S. Department of Agriculture prohibits discrimination in all its programs and activities on the basis of race, color, national origin, gender, religion, age, disability, political beliefs, sexual orientation, and marital or family status.

References

1. Armentano, L.; Bentsáth, A.; Berés, T.; Rusznyák, S.; Szent-Györgyi, A. Über den Einfluss von Substanzen der Flavongruppe auf die Permeabilität der Kapillaren; Vitamin P. *Dtsch. Med. Wochenschr.* **1936**, *62*, 1325–1328.
2. Bentsáth, A.; Rusznyák, S.; Szent-Györgyi, A. Vitamin nature of flavones. *Nature* **1936**, *138*, 798.
3. Rusznyák, S. P.; Szent-Györgyi, A. Vitamin P: Flavonols as vitamins. *Nature* **1936**, *138*, 27.
4. Vickery, H.; Nelson, E.; Almquist, H.; Elvehjem, C. Term "Vitamin P" recommended to be discontinued. *Science* **1950**, *112*, 628.
5. Yao, L. H.; Jiang, Y. M.; Shi, J.; Tomás-Barberán, F. A.; Datta, N.; Singanusong, R.; Chen, S. S. Flavonoids in food and their health benefits. *Plant Foods Hum. Nutr.* **2004**, *59*, 113–122.
6. Manthey, J. A.; Guthrie, N.; Grohmann, K. Biological properties of citrus flavonoids pertaining to cancer and inflammation. *Curr. Med. Chem.* **2001**, *8*, 135–153.
7. Li, S.; Lo, C. Y.; Ho, C. T. Hydroxylated polymethoxyflavones and methylated flavonoids in sweet orange (*Citrus sinensis*) peel. *J. Agric. Food Chem.* **2006**, *54*, 4176–4185.
8. Horowitz, R. M.; Gentili, B. In *Citrus Science and Technology*; Nagy, S., Shaw, P. E., Veldhuis, M. K., Eds.; Avi Publishing Co.: Westport, CN, 1977; Vol 1.
9. Lebreton, M. Sur la matiere crystalline des orangetees, et analyse de ces fruits non encore developpés, famille des Hesperidées. *Société de Pharmacie* **1828**, 377–392.
10. Sawabe, A.; Matsubara, Y. Bioactive glycosides in citrus fruit peels. *Stud. Plant Sci.* **1999**, *6*, 261–274.
11. Freedman, L.; Merritt, A. J. Citrus flavonoid complex: Chemical fractionation and biological activity. *Science* **1963**, *139*, 344–345.
12. Korthuis, R. J.; Gute, D. C. Anti-inflammatory actions of a micronized, purified flavonoid fraction in ischemia/reperfusion. *Adv. Exp. Med. Biol.* **2002**, *505*, 181–190.
13. Bouskela, E.; Donyo, K. A.; Verbeuren, T. J. Effects of Daflon 500 mg on increased microvascular permeability in normal hamsters. *Int J. Microcirc.* **1995**, *14* (suppl. 1), 22–26.
14. Freisnecker, B.; Tsai, A. G.; Intaglietta, M. Cellular basis of inflammation, edema, and the activity of Daflon 500 mg. *Int. J. Microcirc.* **1995**, *15* (suppl. 1), 17–21.

15. Bergan, J. J. Chronic venous insufficiency and the therapeutic effects of Daflon 500 mg. *Angiology* **2005**, *56*, 521–524.
16. Ramelet, A. A. Daflon 500 mg: symptoms and edema: Clinical update. *Angiology* **2005**, *56*, S25–S32.
17. Rizk, S. M.; Sabri, N. A. Evaluation of activity of Daflon in type 2 diabetic patients. *Saudi, Pharm. J.* **2009**, *17*, 199–209.
18. Manthey, J. A.; Grohmann, K.; Montanari, A.; Ash, K.; Manthey, C. L. Polymethoxylated flavones derived from citrus suppress tumor necrosis factor-α expression by human monocytes. *J. Nat. Prod.* **1999**, *62*, 441–444.
19. Gerlo, S.; Verdood, P; Kooijman, R. Modulation of cytokine production by cyclic adenosine monophosphate analogs in human leukocytes. *J. Interferon Cytokine Res.* **2010**, *30*, 883–891.
20. Nikaido, T.; Ohmoto, T.; Sankawa, U.; Tomimori, T.; Miyaichi, Y.; Imoto, Y. Inhibition of adenosine 3′,5′-cyclic monophospate phosphodiesterase by flavonoids. II. *Chem. Pharm. Bull.* **1988**, *36*, 654–661.
21. Petkov, E.; Nokolov, N.; Uzunov, P. Inhibitory effect of some flavonoids and flavonoid mixtures on cyclic AMP phosphodiesterase activity of rat heart. *Planta Med.* **1981**, *43*, 183–186.
22. Manthey, J. A.; Bendele, P. Anti-inflammatory activity of an orange peel polymethoxylated flavone, 3′,4′,3,5,6,7,8-heptamethoxyflavone, in the rat carrageenan/paw edema and mouse lipopolysaccharide-challenge assays. *J. Agric. Food Chem.* **2008**, *56*, 9399–9403.
23. Blackham, A.; Radziwonik, H.; Shaw, I. H. The Arthus reaction in guinea-pig knee joints. A test for anti-inflammatory drugs. *Agents Actions* **1975**, *5*, 519–527.
24. Ungar, G.; Damgaard, E.; Weinstein, H. G. Endocrine control of inflammation: Effect of hormones on anaphylactic arthritis. *Am. J. Physiol.* **1951**, *166*, 340–348.
25. Ungar, G.; Damgaard, E.; Hummel, F. P. Action of salicylates and related drugs on inflammation. *Am. J. Physiol.* **1952**, *171*, 545–553.
26. Manthey, J. A.; Cesar, T. B.; Jackson, E.; Mertens-Talcott, S. Pharmacokinetic study of nobiletin and tangeretin in rat serum by high-performace liquid chromatography-electrospray ionization-mass spectrometry. *J. Agric. Food Chem.* **2011**, *59*, 145–151.

Chapter 11

Functional Role of Walnuts and Açaí Fruits on Brain Health

Shibu M. Poulose* **and Barbara Shukitt-Hale**

USDA-ARS, Human Nutrition Research Center on Aging at Tufts University, Boston, Massachsetts 02111
***E-mail: shibu.poulose@ars.usda.gov.**

Aging is a common risk factor for most diseases, particularly neurodegenerative diseases, which are debilitating in nature and compromise the quality of life, in addition to being economically burdensome. Complete cures for these age-associated brain's diseases are still not in sight; therefore delaying or preventing the pathological manifestation of these diseases through enhancing the brain's endogenous defense mechanisms has gained research attention as an alternate approach. Fruits, nuts and vegetables packed with a multitude of phytochemicals, many of which have potent health benefits, are being actively pursued in this alternate approach. Anthocyanins and polyunsaturated fatty acids have been shown to improve brain function by acting as antioxidant or anti-inflammatory agents as well as altering neuronal signaling mechanisms. Walnuts and Amazonian palm fruit (açaí), are particularly rich in omega fatty acids and anthocyanins respectively, along with many other neuroactive phytochemicals. A review of the functional effects of walnuts and açaí on brain health is presented here.

© 2013 American Chemical Society

Introduction

For survival and endurance, all plants, including fruit- or vegetable-bearing plants, synthesize an enormous number of secondary chemical compounds, widely known as phytochemicals. An abundance of research suggests that many of these phytochemicals extend a multitude of health benefits to humans, mainly through a combination of additive and/or synergistic effects derived from their antioxidant and anti-inflammatory capabilities. On the other hand, aging is largely associated with increased oxidative stress and inflammation in the brain, leading to a cascade of altered signaling and ultimately to the injury or death of neurons. These cascading effects modify neuronal communications, leading to deficits in memory, cognition, motor functions and often to the pathogenesis of neurodegenerative diseases, such as Alzheimer's disease (AD), Parkinson's disease (PD), Huntington's Disease (HD), amyotrophic lateral sclerosis (ALS), prion disease, dementia and so on (*1*, *2*). Apart from being debilitating in nature on the elderly, these diseases cause an enormous economic burden on society with a direct annual economic value for care giving in AD alone estimated at $183 billion in 2011 (*3*). With more than 7 million identified cases of dementia and other neurological disorders in North America and 35.6 million people worldwide living with some form of dementia (*3*), followed by a staggering number of new cases reported on a daily basis, the race for a cure has been greatly intensified. However, it is worth noting that the clinical manifestation of these devastating diseases takes approximately 8-10 years, and in certain cases up to 25 years, indicating a hidden initiation and progression originating in innate brain cells. Etiologies of more than 80% of these incidences indicate a sporadic cause, without any familial-genetic linkage (*1*). Therefore, it has been generally concluded that the pathogenesis for most of the neurological diseases are multifactorial, resulting from a combination of aging, genetic predisposition, and exposure to one or more environmental agents including head trauma, low education level, viruses, and/or toxins, which directly or indirectly increase oxidative stress and inflammation in the brain (*1*, *2*, *4*, *5*). In the search for a cure for these age-related diseases, the emphasis for prevention has gotten much less attention, and therefore it is essential to highlight the alternate nutritional approaches which could potentially boost endogenous defense mechanisms.

Numerous studies from our laboratory and others have reported that feeding lab animals with diets rich in fruits or walnuts has improved memory, cognition and motor functions (*6–10*). These improvements were partly due to the roles of phytochemicals packed in these fruits and walnuts in reducing oxidative stress and inflammation. With such mounting evidence of the beneficial effects of polyphenolics, phytosterols, micronutrients and omega fatty acids, we review here the current knowledge of the phytochemical-rich Amazonian palm fruit (açaí) and walnuts in improving brain health, particularly regarding the molecular events leading to attenuation of oxidative stress and inflammation, and to improving cognition, motor function and other behavioral indices.

Brain Health and Aging: Role of Phytochemicals

The central nervous system is highly susceptible to oxidative stress and inflammation, both increasing together with age. The cost of these insults on the brain is enormous, leading to debilitating neurodegenerative diseases such as AD, PD, ALS, HD, dementia, prion disease, and so on (*11*). Just 2% of the body weight, the brain consumes more than one fifth of the total oxygen and accounts for more than 20% of the metabolic rate (*12*). The balance between the levels of reactive oxygen or nitrogen species generated versus the levels of molecular "quenchers" largely governs the level of damage occurring to brain cells. This intricate balance is often altered with age, resulting in excessive production of reactive species, attenuating countering effects by the endogenous oxido-reductase enzymes and intrinsic proteins. This imbalance leading to increased levels of oxidative stress, inflammation and dysfunction in protein regulation has often been considered the causative factor in declining cognitive abilities such as learning, thinking, decision making, judgment, problem solving and memory (*13*, *14*). The downward spiral of cognitive function with age has been linked to increased risks for various neurodegenerative diseases which immensely compromise the quality of life (*15–17*). While there is a race in search of a cure for these neurodegenerative diseases, so far not much headway has been made. Parallel to therapeutic and alternative medicinal approaches, employing natural compounds from food and herbs, as well as food itself, is being widely considered as a preventive measure to boost the natural defenses of the brain. Numerous studies suggest that consumption of diets rich in fruits, nuts, vegetables, and spices, which are high in antioxidants and anti-inflammatory components, may lower age-related cognitive declines and the risk of developing neurodegenerative disease.

Berries, especially blueberries and strawberries, as well as tree nuts are increasingly popular in this regard, with their immense amount of polyphenolics, polyunsaturated fatty acids and phytosterols. These compounds have contributed towards brain health through significant improvements in cognitive and motor functions. Krikorian et al. reported that daily supplementation of blueberry juice for 12 weeks among older individuals improved cognitive function, particularly with respect to improved paired associate learning, word list recall and reduced depressive symptoms (*18*). Dai et al. (2006) reported that supplementation of fruit and vegetable juice rich in polyphenols at least three times per week attributed to a slower onset of Alzheimer's disease, particularly in patients who are Apolipoprotein-Epsilon4 (ApoE4) carriers (*19*). ApoE4 plays a critical role in promoting amyloid accumulation, neurotoxicity, oxidative stress and neuro fibrillary tangles (*20*) ApoE has been shown to colocalize with amyloid deposits and neurofibrillary tangles in the brain of AD patients. The human ApoE4gene has three alleles E2, E3, and E4, where inheritance of E4 is directly associated with increased risk for AD. In another cohort study of 1,367 subjects above 65 years of age, Commenges and colleagues reported a decreased risk of dementia with increased flavonoid-rich diet consumption (*21*). The Mediterranean diet, consisting of high amounts of fruits, vegetables, cereals, and fish, with minimal amounts of alcohol, red meat and dairy products, has been shown to significantly reduce the risk of AD in a large community-based case-control cohort study involving 194 AD patients and 1790 non-AD patients (*22*). A few other studies also indicate the benefits of the Mediterranean diet in reducing the risk of dementia, as well as mortality, in AD patients, indicating the vital role of fruit and nut bioactive compounds on cognitive health (*23*, *24*).

Walnut Bioactive Components

The potential health benefits of walnuts extend beyond their antioxidant and anti-inflammatory benefits to affect neuronal signaling and inter-neuronal communication. In addition to polyphenols, polyunsaturated fatty acids (PUFA) and phytosterols have been known to contribute to the prevention of age-related neuronal and cognitive decline. While PUFAs are critical in the formation, stability and fluidity of neuronal cell membranes important in synaptic vesicle fusion and neurotransmitter signal transduction (*25*), polyphenols are involved in the restoration of calcium homeostasis in the striatal and hippocampal regions of the brain which are crucial for primary and secondary memory functions (*26*, *27*). Walnuts, containing large amounts of PUFAs and polyphenolics, have been shown to boost brain health and function even with an increase in age (*6*, *28*, *29*). A comprehensive summary of the major phytochemicals present in walnuts is given in Figure 1. Although most of the health benefits are attributed to the PUFAs, walnuts do possess some unique bioactive compounds in fair amounts such as melatonin, which is primarily synthesized by the pineal gland and plays a critical role in regulating circadian rhythms (*30*). Melatonin deficiency has been linked to degeneration of cholinergic neurons in the basal forebrain and the deposition of aggregated proteins, such as amyloid beta peptides (Abeta), leading to cognitive impairment and dementia (*31*). Furthermore, Reiter et al. reported that consumption of walnuts increased blood melatonin level, which was correlated with an increase in "total antioxidant capacity" of the serum, with "total antioxidant capacity" indicating the ability of the blood to detoxify free radicals (*30*). Melatonin is also shown to increase indoleamine levels, thereby protecting against cardiovascular damage and cancer proliferation (*30*). In prevention of tumor growth, the walnut phytochemicals melatonin and eicosapentaenoic acid, an ω-3 fatty acid, are thought to act synergistically by acting as membrane receptors inhibiting the uptake and metabolism of ω-6 linoleic acid, stimulating inhibitory G-proteins, and decreasing the synthesis of intracellular cyclic adenosine monophosphate (cAMP) (*32*). The decrease in cAMP attenuates linoleic acid transport into cancer cells (*32*).

Fatty Acids

Isomer	Name	% Total *
14:0	Myristic acid	0.13
16:0	Palmitic acid	6.70
16:1	Palmitoleic acid	0.23
18:0	Stearic acid	2.27
18:1	Oleic acid	21.0
18:2	Linoleic acid	57.46
18:3	Linolenic acid	11.58
20:0	Arachidic acid	0.08
20:5	Eicosapentaenoic acid	0.06
22:0	Docosanoic acid	0.07

Phytosterols**

Name	mg/kg (kernel)
Total	902-2833
Campesterol	51.0±2.9
Stigmasterol	55.5±11
B-Sitosterol	1129.5±124.6
Cleosterol	11 - 50
Δ^5-Avenasterol	25 -153
Δ^{24}-Stigmastadienol	0-19
Δ^7Avenasterol	0-28

Major essential amino acids

Arginine	Aspartic acid
Glutamic acid	Lysin

Walnuts

Polyphenols (mg/100g) **

Name	Kernel	Pellicle
Hydroxycinnamic acids		
Chlorogenic acid	1-3	24 – 76
Caffeic acid	0.1-0.5	1.8 -11
P-Coumaric acid	0.11-0.30	18-58
Ferrulic acid	0.05-0.11	1.1-6
Sinapic acid	0.02-0.15	1.4-4.0
Hydroxylbenozoic acids		
Syringic acid	17-58	650-1810
Ellagic acid	3.26-10.0	60-270
Gallic acid		
Glansrin		
Juglone	7-20	190-797
Syringaldehyde	0.5-1.15	21-31.5

Micronutrients & Minerals**

Dietary Folate	30 - 40 μg/oz
Dietary Fiber	31-52 g/kg
Melatonin	102 ng/oz
zinc	4.16 mg/oz
Selenium	17 μg/oz
Potassium	> 350 μg/kg
Magnesium	> 300 μg/kg
Phosphorus	> 120 μg/kg

Tocopherols (Vitamin E)**

	μg/g (oil)**
Total	194-350*
α-Tocopherol	51.0±2.9
γ-Tocopherol	55.5±11
δ-Tocopherol	1129.5±124.6

*Oil yield is 51g/100g dry kernel
** Varies with varieties

Figure 1. Major phytochemicals present in walnuts (8, 33–35)

Walnut Effects in Cognition and Signaling

Evidence accumulating from recent studies indicates that walnuts exert specific effects on many biochemical pathways, thereby extending benefits to brain, heart, lungs and other tissues. A walnut diet rich in alpha linolenic acid (ALA) has been shown to reduce inflammatory and cardiovascular risk factors among hypercholesterolemic men and women (*36*). These beneficial effects were attributed to reductions in C-reactive protein (CRP), intracellular adhesion molecules, vascular cell adhesion molecule-1 (VCAM-1) and E-selectin. Similarly, a walnut-rich diet attenuated lipopolysaccharide-(LPS-) induced production of interleukin(IL)-6, IL-1β and TNF-α in human monocytic THP-1 cells and suppressed nuclear factor kappa B- (NF-κB) DNA binding activity (*37*). These effects were also accompanied by activation of peroxisome proliferator-activated receptor gamma (PPARγ). Furthermore, in hypercholesterolemic subjects, walnut diets reduced the production of proinflammatory cytokines (IL-6, IL-1β, TNF-α) in serum (*38*). Reiter and coworkers reported that melatonin, which is synthesized by the pineal gland and which plays a critical role in circadian rhythms, the deficiency of which would cause many biological disorders, was significantly increased in the blood by walnut diets in rodents (*30*). Davis and coworkers further reported that walnut diets extend their heart-healthy benefits via significant reductions in aortic endothelin-1 (ET-1), an important endothelial regulator, in hamsters fed with a high fat diet (*39*). Up to a 75% reduction in ET-1 mRNA levels was observed among the walnut-fed diet group, along with a reduction in aortic cholesterol esters (CE). A mechanism of lowering serum triglycerides by walnuts has also been attributed to the inhibition of stearoyl-CoA desaturase-1 (SCD1), a rate-limiting enzyme responsible for the conversion of saturated fatty acid to monounsaturated fatty acid (*40*). Further studies by the researchers indicated that a reduction in SCD1 resulted in increased cholesterol efflux and decreased intracellular cholesterol storage (*41*). Inhibition of SCD1 is achieved by alpha-linolenic acid through the activation of nuclear receptor farnesoid-X-receptor (FXR), which in turn increases the expression of its target gene, small heterodimer partner (SHP), and decreases liver X-receptor-dependent sterol regulatory element binding protein 1c (SREBP-1c) transcription (*41*). Establishing molecular evidence for the antioxidant effects of a walnut diet among healthy, postmenopausal women, McKay and coworkers reported that supplementation by either 21 or 42 g of raw walnuts/day for 6 weeks significantly increased plasma pyridoxal phosphate (PLP), an active metabolite of vitamin B-6, deficiency of which increases the risk of cardiovascular disease (*42*).

Even though research reports indicating the specific molecular effects in brain by walnuts are notably limited, a few available reports support the beneficial effects of walnuts on brain health. Willis et al. reported that walnut diets enhance cholinergic transmission in aged rats either through increasing acetylcholine (ACh) synthesis or inhibiting acetylcholine esterase, which hydrolyses ACh (*7*). It has been postulated that ACh esterase activity increases with age, leading to dysfunction of cholinergic neurons, in turn leading to cognitive declines similar to those in Alzheimer's disease (*7*). Furthermore, walnut extracts have been shown to attenuate LPS-induced production of nitric oxide (NO) and reduce expression of inducible nitric oxide synthase (iNOS) and TNFα in mouse brain BV2 microglial cells (*43*). Further evaluation of the same study also revealed that walnut extracts internalized the LPS receptor, toll-like receptor-4 (TLR-4), and enhanced antiinflammatory effects through activation of phospholipase D2 (PLD2), an enzyme which hydrolyzes phosphatidylcholine (PC) to produce phosphatidic acid and choline. The health benefits of walnuts in brain were further corroborated by Poulose et al. (2012), where they reported that supplementing

19 month-old rats with walnut diets significantly reduced the aggregation of polyubiquitinated proteins and activated autophagy, a neuronal housekeeping function, in the striatum and hippocampus (*29*). This activation of neuronal housekeeping function is achieved through inhibiting phosphorylation of mTOR, up-regulating ATG5 and Beclin 1 (both ubiquitin-binding proteins), and increasing conversion of MAP1BLC3-I to LC3-II. The clearance of ubiquitinated aggregates such as sequestosome 1 (p62/SQSTM1) was followed by significant reductions in the level of P38-MAP kinase (MAPK) and phosphorylation of nuclear factor kappa B (NFkB) and cyclic AMP responsive element binding protein (CREB) (*29*). All these reports substantially support the hypothesis that the various health benefits of walnuts can in part be attributed to the ability of bioactive components in walnuts to alter signaling at a molecular level. Table 1 summarizes the various major studies reporting the health benefits of walnuts in humans and animals.

Açaí Bioactive Components

Açaí palm fruits, belonging to the genus *Euterpe*, are extensively consumed by the natives of the Amazon delta, who are often found to be healthy into old age, even with limited access to medical care. This graceful aging has infused interest among researchers in regard to the diet of these natives and ultimately to açaí fruits and their bioactive components. There are three primary species of *Euterpe,* i.e., *E. precatoria*, *E. edulis*, and *E. oleracea* (EO), with EO being the most commonly consumed fruit. EO pulp has received considerable attention among food scientists because its antioxidant capacity is very high against all reactive oxygen species, especially the superoxide anion and hydroxyl, peroxyl and peroxynitrite radicals, when compared to other berry fruits. The açaí pulp is rich in many phytochemicals and nutrients which are summarized in Figure 2. In addition to polyphenolics, açaí pulp is very rich in anthocyanins such as cyanidin, delphinidin, malvidin, pelargonidin and peonidin, as well as mono- and poly-unsaturated fatty acids (*66*). Some of the unique phytochemicals in açaí fruit include, but are not limited to, dihydrokaempferol, chrysoerial (a unique flavone), and velutin, a flavonoid with potent anti-inflammatory effects (*67*). Açaí fruits are reported to contain more proteins than an egg, as well as vitamins B1, B2, B3, C, and E (*68*). Being a rich source of omega-3 and omega-6 fatty acids, the oleic acid content of açaí fruit has been reported to be the same as in olive oil (*68*).

Table 1. Summary of the Various Beneficial Effects of Walnuts in Human and Animal Studies

Beneficial Effects	*Study Model*	*References*
Lowers total and low density lipoproteins (LDL)	Healthy Human	(*44, 45*)
Improves serum lipid profile	Hypercholesterolemic Men and Women	(*46–49*)
Inhibition of plasma and LDL oxidation	Human Plasma & LDL	(*50, 51*)
Improves HDL to total cholesterol ratio	Patients with type-2 diabetes	(*52*)
Improves endothelial function	Hypercholesterolemic Men and Women	(*53–55*)
Reduces inflammatory & lipid cardiovascular risk factors	Hypercholesterolemic Men and Women	(*36, 38*)
Helps to achieve optimal fat intake without adverse effects on total fat or energy intake	Patients with type-2 diabetes	(*56*)
Improves postprandial endothelial function (flow mediated dilation)	Hypercholesterolemic patients	(*57*)
Reduces bone resorption	Human	(*58*)
Improves prostate and vascular health in older men	Elderly men	(*59*)
Improves motor and cognitive function, cholinergic transmission	Young adults, Aged rats	(*6–8, 28*)
Inhibition of fibrillation and solubilization of fibrillar amyloid beta protein	Synthetic amyloid beta peptides	(*60*)
Suppresses implanted MDA-MB231 human breast cancer growth	Nude mice	(*61*)
Suppresses mammary gland tumerogenesis and colorectal cancer growth	Transgenic mice	(*62, 63*)
Reduces prostate tumor size and growth, IGF1, resistin and LDL	Transgenic mice	(*64*)
Improves total blood antioxidant capacity	Healthy Human, Rats	(*30, 42, 43, 65*)
Activates neuronal housekeeping	Aged rats	(*29*)

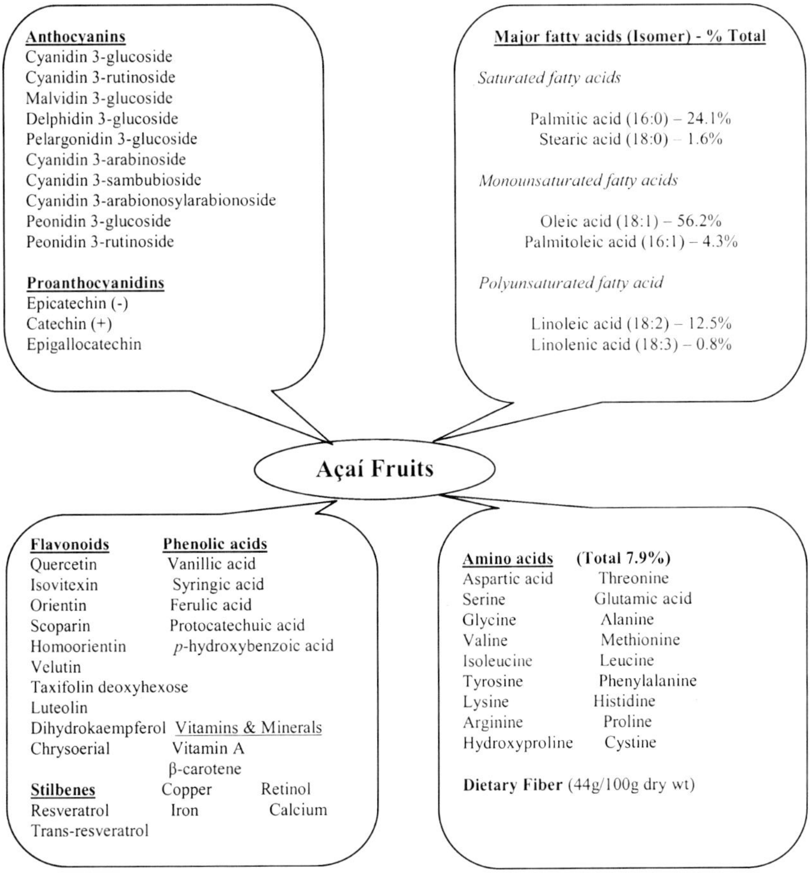

Figure 2. Major phytochemicals present in açaí fruit pulp (66, 67, 69–71).

Health Benefits of Açaí Fruits

Even though human studies examining the effect of açaí fruits on brain health or cognitive function are limited, Jenson et al. reported that giving açaí pulp-rich juice to healthy subjects of 19 to 52 years of age, in a randomized, double-blind, placebo-controlled, crossover study, resulted in significant reductions in lipid peroxidation during oxidative stress, as well as a rapid increase in antioxidant activity in the serum (*72*). In another open-label, clinical pilot study involving human subjects, oral consumption of an açai fruit and berry blend juice was shown to reduce inflammation, improving joint motion and altering pain perception (*73*). Antioxidant and anti-inflammatory effects of açaí have been established in other non-human studies involving either animal models or cell culture experiments. For example, açaí pulp, when added to a high fat diet in studies using the fruit fly (*Drosophila melanogaster)*, has shown to extend the life span via modification of the c-Jun N-terminal kinase (JNK) signaling pathway. In another study by Feio et al. (*74*) , male New Zealand rabbits were fed with a cholesterol-enriched diet for 12 weeks along with açaí pulp supplementation or control. The results indicate that açaí pulp reduced the levels of total cholesterol, non-HDL cholesterol and triglycerides as compared to control. The açaí-fed rabbits had smaller atherosclerotic plaques in their aortas and significantly lower intima/media ratios than controls (*74*). Açaí has also been reported to have analgesic effects in mice, possibly mediated via modulating the levels of prostaglandins (*75*). Açaí stone extracts, which are high in phytosterols, polyunsaturated fatty acids and polyphenols, have been shown to reduce lung inflammation caused by inhalation of cigarette smoke in C57BL/6 mice (*76*). The lungs of mice exposed to cigarette smoke exhibited increases in alveolar macrophage and neutrophil numbers, and myeloperoxidase, superoxide dismutase (SOD), catalase and glutathione peroxidase (GPx) activities compared to mice not exposed to cigarette smoke (*76*). Oral administration of açaí stone extracts was shown to significantly reduce all of the inflammatory indices compared to the control group (*76*). Guerra and coworkers reported that 2% açaí diet supplementation in diabetic rats increased the mRNA levels of γ-glutamylcysteine synthase and glutathione peroxidase in liver tissue and reduced the production of reactive oxygen species by neutrophils, thiobarbituric acid-reactive substances and carbonyl proteins in the hepatic tissues (*77*). Anti-inflammatory effects of açaí extracts have been further established by isolating velutin, a rare flavonoid present in only certain fruits, by Xie and coworkers (*67*). In this study, velutin was shown to reduce the lipopolysaccaride-induced production of TNF-α and IL-6 via inhibiting the transcription factor viz. NF-κB in mouse peripheral macrophages as well as peritoneal macrophages (*67*). Furthermore, velutin was shown to reduce the phosphorylation of p38-mitogen activated protein kinase (MAPK) and JNK (*67*).

To date, studies exploring the health effects of açaí fruits on humans, particularly on brain health, in terms of cognition and motor functions, have been extremely limited. Numerous studies suggest that declines in cognition may be the result of an increasing sensitivity to oxidative and inflammatory stressors that occurs during the aging process. Supplementary evidence has suggested that strong antioxidant and anti-inflammatory effects by açaí fruits could provide possible protection to brain cells in vivo, but this is yet to be determined. In a similar context, we recently reported that potential health benefits by anthocyanin-rich açaí pulp extracts extend beyond their antioxidant and anti-inflammatory properties to affect neuronal signaling involving several oxidative stress and inflammatory mediators in mouse brain microglial cells (*66*). In this study, BV-2 microglial cells pretreated with different fractions of açai pulp extract showed significantly reduced nitrite production induced by LPS, which was correlated with a decrease in inducible nitric oxide synthase (iNOS) expression. The protection of microglial cells by açai pulp fractions was also accompanied by significant concentration-dependent reductions in cyclooxygenase-2 (COX-2), p38 mitogen-activated protein kinase (p38-MAPK), tumor necrosis factor-alpha (TNF-α) and nuclear factor kappa-b (NF-κB) (*66*). In another study using animal brain tissues, Spada and coworkers reported that pretreatment with açaí pulp attenuated hydrogen peroxide-induced damage of both lipids and proteins in the cerebral cortex, hippocampus and cerebellum of adult rats (*78*). There are numerous studies involving isolated phytochemicals from other berry sources, which are also found in açaí fruits. For example, aged mice fed with the flavonoid luteolin for 8-12 weeks have shown suppression of proinflammatory microglial activation and improved hippocampal-mediated spatial and working memory (*79*). Under the context of a limited number of studies on brain health involving açaí fruits, it is hard to specify the definitive effects of açaí on cognitive function, and efforts are underway to determine such benefits through research involving more animal studies and potential human studies.

Conclusions

Aging has become increasingly recognized as a common risk factor for most diseases, and more so for the debilitating diseases involving the brain (Alzheimer's, Parkinson's, dementia, and others). An increase in age firmly accompanies a decreased ability of the brain to counter stresses, particularly oxidative stress and inflammation both of which often cause a multitude of functional losses, including the degeneration or death of neurons, apart from being implicated in an array of diseases including cancers and cardiovascular disease. Dietary supplementation with walnuts or açaí extracts, which are high in numerous phytochemicals, can decrease this vulnerability to oxidative stress and inflammation. Antioxidant-rich diet supplementation has been shown to retard and even reverse age-related declines in brain function, and in cognitive and motor performance, in animal and human studies. Therefore, nutritional interventions with diets rich in nuts, berries, fruits and vegetables in an aging population may reverse or allay age-related motor or cognitive deficits, delay the onset of age-related neurodegenerative diseases and convey long-term health dividends.

References

1. Marx, J. Neurodegenerative diseases. Picking apart the causes of mysterious dementias. *Science* **2006**, *314* (5796), 42–43.
2. Floyd, R. A.; Hensley, K. Oxidative stress in brain aging. Implications for therapeutics of neurodegenerative diseases. *Neurobiol. Aging* **2002**, *23* (5), 795–807.
3. Thies, W.; Bleiler, L. 2011 Alzheimer's disease facts and figures. *Alzheimer's Dementia* **2011**, *7* (2), 208–244.
4. Hardy, J. The genetic causes of neurodegenerative diseases. *J. Alzheimer's Dis.* **2001**, *3* (1), 109–116.
5. Pierpaoli, W. Neurodegenerative diseases: A common etiology and a common therapy. *Ann N. Y. Acad. Sci.* **2005**, *1057*, 319–326.
6. Willis, L. M.; Shukitt-Hale, B.; Joseph, J. A. Modulation of cognition and behavior in aged animals: Role for antioxidant- and essential fatty acid-rich plant foods. *Am. J. Clin. Nutr.* **2009**, *89* (5), 1602S–1606S.

7. Willis, L. M.; Shukitt-Hale, B.; Joseph, J. A. Dietary polyunsaturated fatty acids improve cholinergic transmission in the aged brain. *Genes Nutr.* **2009**, *4* (4), 309–314.
8. Willis, L. M.; Shukitt-Hale, B.; Cheng, V.; Joseph, J. A. Dose-dependent effects of walnuts on motor and cognitive function in aged rats. *Br. J. Nutr.* **2009**, *101* (8), 1140–1144.
9. Shukitt-Hale, B.; Cheng, V.; Bielinski, D. F.; Joseph, J. A. Walnuts can improve motor and cognitive function in aged rats. *Soc. Neurosci. Abs.* **2007**, *33*, 256.14.
10. Shukitt-Hale, B.; Lau, F. C.; Joseph, J. A. Berry fruit supplementation and the aging brain. *J. Agric. Food Chem.* **2008**, *56* (3), 636–641.
11. Emerit, J.; Edeas, M.; Bricaire, F. Neurodegenerative diseases and oxidative stress. *Biomed. Pharmacother.* **2004**, *58* (1), 39–46.
12. Magistretti, P. J.; Pellerin, L.; Martin, J.-L. *Brain Energy Metabolism: An Integrated Cellular Perspective*; Raven Press: New York, 1995.
13. Joseph, J. A.; Shukitt-Hale, B.; Denisova, N. A.; Bielinski, D.; Martin, A.; McEwen, J. J.; Bickford, P. C. Reversals of age-related declines in neuronal signal transduction, cognitive, and motor behavioral deficits with blueberry, spinach, or strawberry dietary supplementation. *J. Neurosci.* **1999**, *19* (18), 8114–8121.
14. Shukitt-Hale, B.; McEwen, J. J.; Szprengiel, A.; Joseph, J. A. Effect of age on the radial arm water maze-a test of spatial learning and memory. *Neurobiol. Aging* **2004**, *25* (2), 223–229.
15. Douglas, P. M.; Dillin, A. Protein homeostasis and aging in neurodegeneration. *J. Cell Biol.* **2010**, *190* (5), 719–729.
16. Joseph, J.; Cole, G.; Head, E.; Ingram, D. Nutrition, brain aging, and neurodegeneration. *J. Neurosci.* **2009**, *29* (41), 12795–12801.
17. Joseph, J. A.; Shukitt-Hale, B.; Casadesus, G. Reversing the deleterious effects of aging on neuronal communication and behavior: Beneficial properties of fruit polyphenolic compounds. *Am. J. Clin. Nutr.* **2005**, *81* (Suppl 1), 313S–316S.
18. Krikorian, R.; Shidler, M. D.; Nash, T. A.; Kalt, W.; Vinqvist-Tymchuk, M. R.; Shukitt-Hale, B.; Joseph, J. A. Blueberry supplementation improves memory in older adults. *J. Agric. Food Chem.* **2010**, *58* (7), 3996–4000.
19. Dai, Q.; Borenstein, A. R.; Wu, Y.; Jackson, J. C.; Larson, E. B. Fruit and vegetable juices and Alzheimer's disease: The Kame Project. *Am. J. Med.* **2006**, *119* (9), 751–759.
20. Lahiri, D. K. Apolipoprotein E as a target for developing new therapeutics for Alzheimer's disease based on studies from protein, RNA, and regulatory region of the gene. *J. Mol. Neurosci.* **2004**, *23* (3), 225–233.
21. Commenges, D.; Scotet, V.; Renaud, S.; Jacqmin-Gadda, H.; Barberger-Gateau, P.; Dartigues, J. F. Intake of flavonoids and risk of dementia. *Eur. J. Epidemiol.* **2000**, *16* (4), 357–363.
22. Morris, M. C.; Evans, D. A.; Bienias, J. L.; Tangney, C. C.; Bennett, D. A.; Aggarwal, N.; Wilson, R. S.; Scherr, P. A. Dietary intake of antioxidant nutrients and the risk of incident Alzheimer disease in a biracial community study. *JAMA, J. Am. Med. Assoc.* **2002**, *287* (24), 3230–3237.

23. Scarmeas, N.; Stern, Y.; Tang, M. X.; Mayeux, R.; Luchsinger, J. A. Mediterranean diet and risk for Alzheimer's disease. *Ann. Neurol.* **2006**, *59* (6), 912–921.
24. Damasceno, N. R.; Perez-Heras, A.; Serra, M.; Cofan, M.; Sala-Vila, A.; Salas-Salvado, J.; Ros, E. Crossover study of diets enriched with virgin olive oil, walnuts or almonds. Effects on lipids and other cardiovascular risk markers. *Nutr., Metab. Cardiovasc. Dis.* **2011**, *21* (Suppl 1), S14–S20.
25. Joseph, J. A.; Shukitt-Hale, B.; Willis, L. M. *Mitigation of Oxidative Stress and Inflammatory Signalling by Fruit and Walnut Polyphenols: Implications for Cognitive Aging*; Wiley-Blackwell: Hoboken, NJ, 2010; Vol. 2, pp 283–291.
26. Devan, B. D.; Goad, E. H.; Petri, H. L. Dissociation of hippocampal and striatal contributions to spatial navigation in the water maze. *Neurobiol. Learn. Mem.* **1996**, *66* (3), 305–323.
27. Oliveira, M. G. M.; Bueno, O. F. A.; Pomarico, A. C.; Gugliano, E. B. Strategies used by hippocampal- and caudate-putamen-lesioned rats in a learning task. *Neurobiol. Learn. Mem.* **1997**, *68* (1), 32–41.
28. Pribis, P.; Bailey, R. N.; Russell, A. A.; Kilsby, M. A.; Hernandez, M.; Craig, W. J.; Grajales, T.; Shavlik, D. J.; Sabate, J. Effects of walnut consumption on cognitive performance in young adults. *Br. J. Nutr.* **2011**, 1–9.
29. Poulose, S. M.; Bielinski, D. F.; Shukitt-Hale, B. Walnut diet reduces accumulation of polyubiquitinated proteins and inflammation in the brain of aged rats. *J. Nutr. Biochem.* **2012**, DOI: 10.1016/j.bbr.2011.03.031.
30. Reiter, R. J.; Manchester, L. C.; Tan, D. X. Melatonin in walnuts: Influence on levels of melatonin and total antioxidant capacity of blood. *Nutrition* **2005**, *21* (9), 920–924.
31. Lahiri, D. K.; Chen, D. M.; Lahiri, P.; Bondy, S.; Greig, N. H. Amyloid, cholinesterase, melatonin, and metals and their roles in aging and neurodegenerative diseases. *Ann. N. Y. Acad. Sci.* **2005**, *1056*, 430–449.
32. Sauer, L. A.; Dauchy, R. T.; Blask, D. E. Polyunsaturated fatty acids, melatonin, and cancer prevention. *Biochem. Pharmacol.* **2001**, *61* (12), 1455–1462.
33. Pereira, J. A.; Oliveira, I.; Sousa, A.; Valentao, P.; Andrade, P. B.; Ferreira, I. C.; Ferreres, F.; Bento, A.; Seabra, R.; Estevinho, L. Walnut (*Juglans regia* L.) leaves: Phenolic compounds, antibacterial activity and antioxidant potential of different cultivars. *Food Chem. Toxicol.* **2007**, *45* (11), 2287–2295.
34. Oliveira, I.; Sousa, A.; Ferreira, I. C.; Bento, A.; Estevinho, L.; Pereira, J. A. Total phenols, antioxidant potential and antimicrobial activity of walnut (*Juglans regia* L.) green husks. *Food Chem. Toxicol.* **2008**, *46* (7), 2326–2331.
35. Pereira, J. A.; Oliveira, I.; Sousa, A.; Ferreira, I. C.; Bento, A.; Estevinho, L. Bioactive properties and chemical composition of six walnut (*Juglans regia* L.) cultivars. *Food Chem. Toxicol.* **2008**, *46* (6), 2103–2111.
36. Zhao, G.; Etherton, T. D.; Martin, K. R.; West, S. G.; Gillies, P. J.; Kris-Etherton, P. M. Dietary alpha-linolenic acid reduces inflammatory and lipid

cardiovascular risk factors in hypercholesterolemic men and women. *J. Nutr.* **2004**, *134* (11), 2991–2997.

37. Zhao, G.; Etherton, T. D.; Martin, K. R.; Vanden Heuvel, J. P.; Gillies, P. J.; West, S. G.; Kris-Etherton, P. M. Anti-inflammatory effects of polyunsaturated fatty acids in THP-1 cells. *Biochem. Biophys. Res. Commun.* **2005**, *336* (3), 909–917.
38. Zhao, G.; Etherton, T. D.; Martin, K. R.; Gillies, P. J.; West, S. G.; Kris-Etherton, P. M. Dietary alpha-linolenic acid inhibits proinflammatory cytokine production by peripheral blood mononuclear cells in hypercholesterolemic subjects. *Am. J. Clin. Nutr.* **2007**, *85* (2), 385–391.
39. Davis, P.; Valacchi, G.; Pagnin, E.; Shao, Q.; Gross, H. B.; Calo, L.; Yokoyama, W. Walnuts reduce aortic ET-1 mRNA levels in hamsters fed a high-fat, atherogenic diet. *J. Nutr.* **2006**, *136* (2), 428–432.
40. Velliquette, R. A.; Gillies, P. J.; Kris-Etherton, P. M.; Green, J. W.; Zhao, G.; Vanden Heuvel, J. P. Regulation of human stearoyl-CoA desaturase by omega-3 and omega-6 fatty acids: Implications for the dietary management of elevated serum triglycerides. *J. Clin. Lipidol.* **2009**, *3* (4), 281–288.
41. Zhang, J.; Kris-Etherton, P. M.; Thompson, J. T.; Hannon, D. B.; Gillies, P. J.; Vanden Heuvel, J. P. Alpha-linolenic acid increases cholesterol efflux in macrophage-derived foam cells by decreasing stearoyl CoA desaturase 1 expression: Evidence for a farnesoid-X-receptor mechanism of action. *J. Nutr. Biochem.* **2012**, *23* (4), 400–409.
42. McKay, D. L.; Chen, C. Y.; Yeum, K. J.; Matthan, N. R.; Lichtenstein, A. H.; Blumberg, J. B. Chronic and acute effects of walnuts on antioxidant capacity and nutritional status in humans: A randomized, cross-over pilot study. *Nutr. J.* **2010**, *9*, 21.
43. Willis, L. M.; Bielinski, D. F.; Fisher, D. R.; Matthan, N. R.; Joseph, J. A. Walnut extract inhibits LPS-induced activation of BV-2 microglia via internalization of TLR4: Possible involvement of phospholipase D2. *Inflammation* **2010**, *33* (5), 325–333.
44. Sabate, J.; Fraser, G. E.; Burke, K.; Knutsen, S. F.; Bennett, H.; Lindsted, K. D. Effects of walnuts on serum lipid levels and blood pressure in normal men. *N. Engl. J. Med.* **1993**, *328* (9), 603–7.
45. Almario, R. U.; Vonghavaravat, V.; Wong, R.; Kasim-Karakas, S. E. Effects of walnut consumption on plasma fatty acids and lipoproteins in combined hyperlipidemia. *Am. J. Clin. Nutr.* **2001**, *74* (1), 72–79.
46. Zambon, D.; Sabate, J.; Munoz, S.; Campero, B.; Casals, E.; Merlos, M.; Laguna, J. C.; Ros, E. Substituting walnuts for monounsaturated fat improves the serum lipid profile of hypercholesterolemic men and women. A randomized crossover trial. *Ann. Intern. Med.* **2000**, *132* (7), 538–546.
47. Munoz, S.; Merlos, M.; Zambon, D.; Rodriguez, C.; Sabate, J.; Ros, E.; Laguna, J. C. Walnut-enriched diet increases the association of LDL from hypercholesterolemic men with human HepG2 cells. *J. Lipid. Res.* **2001**, *42* (12), 2069–2076.
48. Banel, D. K.; Hu, F. B. Effects of walnut consumption on blood lipids and other cardiovascular risk factors: A meta-analysis and systematic review. *Am. J. Clin. Nutr.* **2009**, *90* (1), 56–63.

49. Iwamoto, M.; Imaizumi, K.; Sato, M.; Hirooka, Y.; Sakai, K.; Takeshita, A.; Kono, M. Serum lipid profiles in Japanese women and men during consumption of walnuts. *Eur. J. Clin. Nutr.* **2002**, *56* (7), 629–637.
50. Feldman, E. B. The scientific evidence for a beneficial health relationship between walnuts and coronary heart disease. *J. Nutr.* **2002**, *132* (5), 1062S–1101S.
51. Torabian, S.; Haddad, E.; Rajaram, S.; Banta, J.; Sabate, J. Acute effect of nut consumption on plasma total polyphenols, antioxidant capacity and lipid peroxidation. *J. Hum. Nutr. Diet* **2009**, *22* (1), 64–71.
52. Tapsell, L. C.; Gillen, L. J.; Patch, C. S.; Batterham, M.; Owen, A.; Bare, M.; Kennedy, M. Including walnuts in a low-fat/modified-fat diet improves HDL cholesterol-to-total cholesterol ratios in patients with type 2 diabetes. *Diabetes Care* **2004**, *27* (12), 2777–2783.
53. Ros, E.; Nunez, I.; Perez-Heras, A.; Serra, M.; Gilabert, R.; Casals, E.; Deulofeu, R. A walnut diet improves endothelial function in hypercholesterolemic subjects: A randomized crossover trial. *Circulation* **2004**, *109* (13), 1609–1614.
54. Aronis, K. N.; Vamvini, M. T.; Chamberland, J. P.; Sweeney, L. L.; Brennan, A. M.; Magkos, F.; Mantzoros, C. S. Short-term walnut consumption increases circulating total adiponectin and apolipoprotein A concentrations, but does not affect markers of inflammation or vascular injury in obese humans with the metabolic syndrome: Data from a double-blinded, randomized, placebo-controlled study. *Metabolism* **2012**, *61* (4), 577–582.
55. West, S. G.; Krick, A. L.; Klein, L. C.; Zhao, G.; Wojtowicz, T. F.; McGuiness, M.; Bagshaw, D. M.; Wagner, P.; Ceballos, R. M.; Holub, B. J.; Kris-Etherton, P. M. Effects of diets high in walnuts and flax oil on hemodynamic responses to stress and vascular endothelial function. *J. Am. Coll. Nutr.* **2010**, *29* (6), 595–603.
56. Gillen, L. J.; Tapsell, L. C.; Patch, C. S.; Owen, A.; Batterham, M. Structured dietary advice incorporating walnuts achieves optimal fat and energy balance in patients with type 2 diabetes mellitus. *J. Am. Diet. Assoc.* **2005**, *105* (7), 1087–1096.
57. Cortes, B.; Nunez, I.; Cofan, M.; Gilabert, R.; Perez-Heras, A.; Casals, E.; Deulofeu, R.; Ros, E. Acute effects of high-fat meals enriched with walnuts or olive oil on postprandial endothelial function. *J. Am. Coll. Cardiol.* **2006**, *48* (8), 1666–16671.
58. Griel, A. E.; Kris-Etherton, P. M.; Hilpert, K. F.; Zhao, G.; West, S. G.; Corwin, R. L. An increase in dietary n-3 fatty acids decreases a marker of bone resorption in humans. *Nutr. J.* **2007**, *6*, 2.
59. Spaccarotella, K. J.; Kris-Etherton, P. M.; Stone, W. L.; Bagshaw, D. M.; Fishell, V. K.; West, S. G.; Lawrence, F. R.; Hartman, T. J. The effect of walnut intake on factors related to prostate and vascular health in older men. *Nutr. J.* **2008**, *7*, 13.
60. Chauhan, N.; Wang, K. C.; Wegiel, J.; Malik, M. N. Walnut extract inhibits the fibrillization of amyloid beta-protein, and also defibrillizes its preformed fibrils. *Curr. Alzheimer Res.* **2004**, *1* (3), 183–188.

61. Hardman, W. E.; Ion, G. Suppression of implanted MDA-MB 231 human breast cancer growth in nude mice by dietary walnut. *Nutr. Cancer* **2008**, *60* (5), 666–674.
62. Hardman, W. E.; Ion, G.; Akinsete, J. A.; Witte, T. R. Dietary walnut suppressed mammary gland tumorigenesis in the C(3)1 TAg mouse. *Nutr. Cancer* **2011**, *63* (6), 960–970.
63. Nagel, J. M.; Brinkoetter, M.; Magkos, F.; Liu, X.; Chamberland, J. P.; Shah, S.; Zhou, J.; Blackburn, G.; Mantzoros, C. S. Dietary walnuts inhibit colorectal cancer growth in mice by suppressing angiogenesis. *Nutrition* **2012**, *28* (1), 67–75.
64. Davis, P. A.; Vasu, V. T.; Gohil, K.; Kim, H.; Khan, I. H.; Cross, C. E.; Yokoyama, W. A high-fat diet containing whole walnuts (*Juglans regia*) reduces tumour size and growth along with plasma insulin-like growth factor 1 in the transgenic adenocarcinoma of the mouse prostate model. *Br. J. Nutr.* **2012**, *108* (10), 1764–1772.
65. Qamar, W.; Sultana, S. Polyphenols from Juglans regia L. (walnut) kernel modulate cigarette smoke extract induced acute inflammation, oxidative stress and lung injury in Wistar rats. *Hum. Exp. Toxicol.* **2011**, *30* (6), 499–506.
66. Poulose, S. M.; Fisher, D. R.; Larson, J. A.; Bielinski, D. F.; Rimando, A. M.; Carey, A. N.; Schauss, A. G.; Shukitt-Hale, B. Anthocyanin-rich açaí (*Euterpe oleracea* Mart.) fruit pulp fractions attenuate inflammatory stress signaling in mouse brain BV-2 microglial cells. *J. Agric. Food Chem.* **2012**, *60* (4), 1084–1093.
67. Xie, C.; Kang, J.; Li, Z.; Schauss, A. G.; Badger, T. M.; Nagarajan, S.; Wu, T.; Wu, X. The açaí flavonoid velutin is a potent anti-inflammatory agent: Blockade of LPS-mediated TNF-alpha and IL-6 production through inhibiting NF-kappaB activation and MAPK pathway. *J. Nutr. Biochem.* **2012**, *23* (9), 1184–1191.
68. Ulbricht, C.; Brigham, A.; Burke, D.; Costa, D.; Giese, N.; Iovin, R.; Grimes Serrano, J. M.; Tanguay-Colucci, S.; Weissner, W.; Windsor, R. An evidence-based systematic review of açaí (*Euterpe oleracea*) by the Natural Standard Research Collaboration. *J. Diet. Suppl.* **2012**, *9* (2), 128–147.
69. Pacheco-Palencia, L. A.; Mertens-Talcott, S.; Talcott, S. T. Chemical composition, antioxidant properties, and thermal stability of a phytochemical enriched oil from açaí (*Euterpe oleracea* Mart.). *J. Agric. Food Chem.* **2008**, *56* (12), 4631–4636.
70. Schauss, A. G.; Wu, X.; Prior, R. L.; Ou, B.; Patel, D.; Huang, D.; Kababick, J. P. Phytochemical and nutrient composition of the freeze-dried amazonian palm berry, *Euterpe oleraceae* mart. (açaí). *J. Agric. Food Chem.* **2006**, *54* (22), 8598–8603.
71. Del Pozo-Insfran, D.; Percival, S. S.; Talcott, S. T. Açaí (*Euterpe oleracea* Mart.) polyphenolics in their glycoside and aglycone forms induce apoptosis of HL-60 leukemia cells. *J. Agric. Food Chem.* **2006**, *54* (4), 1222–1229.
72. Jensen, G. S.; Wu, X.; Patterson, K. M.; Barnes, J.; Carter, S. G.; Scherwitz, L.; Beaman, R.; Endres, J. R.; Schauss, A. G. In vitro and in vivo antioxidant and anti-inflammatory capacities of an antioxidant-rich fruit

and berry juice blend. Results of a pilot and randomized, double-blinded, placebo-controlled, crossover study. *J. Agric. Food Chem.* **2008**, *56* (18), 8326–8333.

73. Jensen, G. S.; Ager, D. M.; Redman, K. A.; Mitzner, M. A.; Benson, K. F.; Schauss, A. G. Pain reduction and improvement in range of motion after daily consumption of an açaí (*Euterpe oleracea* Mart.) pulp-fortified polyphenolic-rich fruit and berry juice blend. *J. Med. Food* **2011**, *14* (7–8), 702–711.
74. Feio, C. A.; Izar, M. C.; Ihara, S. S.; Kasmas, S. H.; Martins, C. M.; Feio, M. N.; Maues, L. A.; Borges, N. C.; Moreno, R. A.; Povoa, R. M.; Fonseca, F. A. *Euterpe oleracea* (açaí) modifies sterol metabolism and attenuates experimentally-induced atherosclerosis. *J. Atheroscler. Thromb.* **2012**, *19* (3), 237–245.
75. Ribeiro, J. C.; Antunes, L. M.; Aissa, A. F.; Darin, J. D.; De Rosso, V. V.; Mercadante, A. Z.; Bianchi Mde, L. Evaluation of the genotoxic and antigenotoxic effects after acute and subacute treatments with açaí pulp (*Euterpe oleracea* Mart.) on mice using the erythrocytes micronucleus test and the comet assay. *Mutat. Res.* **2010**, *695* (1–2), 22–28.
76. de Moura, R. S.; Ferreira, T. S.; Lopes, A. A.; Pires, K. M.; Nesi, R. T.; Resende, A. C.; Souza, P. J.; da Silva, A. J.; Borges, R. M.; Porto, L. C.; Valenca, S. S. Effects of *Euterpe oleracea* Mart. (ACAI) extract in acute lung inflammation induced by cigarette smoke in the mouse. *Phytomedicine* **2012**, *19* (3–4), 262–269.
77. Guerra, J. F.; Magalhaes, C. L.; Costa, D. C.; Silva, M. E.; Pedrosa, M. L. Dietary açaí modulates ROS production by neutrophils and gene expression of liver antioxidant enzymes in rats. *J. Clin. Biochem. Nutr.* **2011**, *49* (3), 188–194.
78. Spada, P. D.; Dani, C.; Bortolini, G. V.; Funchal, C.; Henriques, J. A.; Salvador, M. Frozen fruit pulp of *Euterpe oleraceae* Mart. (açaí) prevents hydrogen peroxide-induced damage in the cerebral cortex, cerebellum, and hippocampus of rats. *J. Med. Food* **2009**, *12* (5), 1084–1088.
79. Jang, S.; Dilger, R. N.; Johnson, R. W. Luteolin inhibits microglia and alters hippocampal-dependent spatial working memory in aged mice. *J. Nutr.* **2010**, *140* (10), 1892–1898.

Chapter 12

Seeking a New Anti-Skin-Aging Material: Piceatannol and Its Derivatives from Passion Fruit *(Passiflora edulis)* Seed

Yuko Matsui,*,† Kenkichi Sugiyama,† Masanori Kamei,† Toshio Takahashi,† Tamio Suzuki,‡ Yohtaro Katagata,§ and Tatsuhiko Ito#

†Research Institute, Morinaga & Co., Ltd., Japan
‡Department of Dermatology, Yamagata University School of Medicine, Yamagata University, Japan
§Faculty of Agriculture and Life Science, Hirosaki University, Japan
#Health Care Division, Morinaga & Co., Ltd., Japan
***E-mail: y-matsui-jd@morinaga.co.jp. Tel: +81-45-571-2504. Fax: +81-45-571-6109. Address: Research Institute, Morinaga & Co., Ltd., 2-1-1, Shimosueyoshi, Tsurumi-ku, Yokohama 230-8504, Japan.**

Passion fruit, the fruit of *Passiflora edulis*, is grown in tropical regions, such as South America and South Asia. The presence of a large amount of polyphenols have been discovered in the seed of passion fruit with piceatannol (3,4,3′,5′-tetrahydroxy-*trans*-stilbene) being the major constituent. Derivatives of piceatannol were also detected. We have focused on the effect of passion fruit on skin health using cultured human melanoma and fibroblast cells, to investigate the inhibition of melanogenesis and promotion of collagen synthesis. Positive effects were obtained. Piceatannol and its derivatives were found to be responsible for the effects observed for passion fruit seed extracts.

© 2013 American Chemical Society

Introduction

Tropical and subtropical fruits and vegetables are beneficial for human health, and the interest in these foods is expanding, not only among the scientific community, but also among the consumers who purchase products containing these fruits and vegetables. Açaí (*Euterpe oleracea*), mango (*Mangifera indica*), pomegranate (*Punica granatum*), and many tropical fruits have attracted the attention of consumers since the word "superfruit" was coined. Their inclusion in fresh juices, cereals, snacks, jams, and many other foods is appealing because of their antioxidant effects. The high content of polyphenols in various tropical fruits (*1*) and the bioactive effects of tropical fruits, such as açaí and mangosteen (*2*, *3*), have been reported.

Passion fruit, the fruit of *Passiflora edulis* from the Passifloraceae family, is a well-known tropical plant from the South American tropical forests. Passion fruit is usually consumed as juice or eaten in its natural state. The leaves, vines, and flowers of members of the Passifloraceae family have been used as medicinal herbs. There are many studies on the actions of Passifloraceae herbs, which include their antianxiety effect in humans (*4*) and their anti-inflammatory and cough-suppressant effects in mice (*5*, *6*). Research on passion fruit are limited, however, studies on the phytochemical composition such as polyphenolic compounds (*7*), molecules in the carotenoid family (*8*), and vitamin C contents (*9*) are reported.

As mentioned above, the interest in polyphenols in such fruit is increasing. Polyphenols in fruit became well-known since the report that shows the moderate consumption of red wine reduces the risk of cardiovascular disease has spread abroad (*10*). Resveratrol became one of the popular polyphenols, which is contained in grapes (*11*). The content in different grape cultivars (*12*) as well as the bioactivity of resveratrol is well studied which includes antioxidant activity (*13*), anti- cancer effects (*14*) and so on.

The health effects of the polyphenols contained in many natural plants, including tropical fruits, have been increasingly and energetically studied, and a large number of them have focused on skin health. The market of food and cosmetic products that targets anti-skin-aging effects is expanding. Skin abnormalities are caused by ultraviolet (UV) exposure, mental and environmental stress, eating habits, etc., leading to skin pigmentation, wrinkles, and even skin cancer. Passion fruit contains many kinds of phytochemicals as mentioned before, and they are known to be beneficial for the skin. Many bioactive substances from naturally occurring plants have been studied regarding the prevention of melanogenesis. Those reports describe the effect of polyphenols from safflower seed, grape seed (*15*, *16*), and many other edible plants. Collagen, which is another molecule that is essential for skin health, plays many important roles in the body, including cell-to-cell adhesion, cell proliferation, and cell differentiation. The functional properties of the skin depend considerably on the quantity and condition of the collagen present in the dermis. Some foods and food components, e.g., royal jelly (*17*) and polyphenols (such as catechin and flavonoids) (*18*, *19*), contribute to the maintenance of collagen condition in the skin or inhibit collagen-degrading factors. To prevent skin damage and maintain its protective potency against environmental agents, an increasing number of research projects have focused on compounds that target the skin. As described above, polyphenols play important roles in dermal cells; thus, we hypothesized the effectiveness of passion fruit in the promotion and maintenance of skin health.

Experimental

Materials

Piceatannol (3,4,3′,5′-tetrahydroxy-*trans*-stilbene) and resveratrol (3,5,4′-trihydroxy-*trans*-stilbene) were purchased from Sigma-Aldrich Japan (Tokyo, Japan). Culture reagents, such as Dulbecco's modified Eagle's medium (DMEM), Medium 106S, phosphate-buffered saline (PBS), low-serum growth supplement (LSGS), glutamine, penicillin, streptomycin, gentamicin, amphotericin B, and trypsin-EDTA, were purchased from Invitrogen (Carlsbad, CA). Fetal bovine serum (FBS) was purchased from Biowest (Maine-et-Loire, France). Synthetic melanin was purchased from Nacalai Tesque (Kyoto, Japan) and the Sircol™ soluble collagen assay kit was purchased from Biocolor Life Science Assays (Newtownabbey, UK). Other reagents were purchased from Wako Pure Chemicals (Tokyo, Japan).

Polyphenol Quantification

The rind, pulp, and seeds of passion fruit (product of Kagoshima, Japan) were separated from whole passion fruit and freeze-dried, milled, and extracted twice using 10-fold amounts of 80% (v/v) ethanol by shaking at room temperature. The extract solution was centrifuged at 3,000 × g for 10 min to obtain the supernatant, which was filtered through a paper filter to remove the sediment. Aliquots containing the extracts were evaporated, freeze-dried, and dissolved in distilled water at the concentration of 20 mg/mL. The extract solutions were filtered using a hydrophilic membrane filter (Kanto Kagaku, Tokyo, Japan) with a pore size of 0.45 μm.

The concentration of polyphenols in the extracts of rind, pulp, and seeds was measured according to a modified Folin–Ciocalteu method, as described previously (*20*), using (–)-epicatechin to obtain the standard curve for polyphenol.

Qualitative Analysis and Determination of Piceatannol

Ground passion fruit seeds were extracted with 70% (v/v) acetone three times, with shaking at room temperature. Samples were evaporated and freeze-dried to obtain crude extracts. A total of 100 mg of crude extract was suspended in 50 mL of a 50% methanol solution and centrifuged at 1500 × *g* for 5 min. The supernatant was separated via reverse-phase liquid chromatography using a linear-gradient mode, as follows. Chromatographic measurements were carried out using an Agilent 1100 Series liquid chromatography mass spectrometry (LC/MS) system (Agilent Technologies, Tokyo, Japan) that included a photodiode array (PDA) detector. The high-performance liquid chromatography (HPLC) column used in this study was Inertsil octa decyl silyl (ODS) -3 HPLC column (150 mm × 2.1 mm inner diameter, 5 μm; GL Science, Tokyo, Japan). The mobile phase consisted of (A) water and (B) acetonitrile (v/v) using an initial gradient elution of 10% (B) and a gradient of 10–45% (B) at 0–25 min. The column temperature was maintained at 45 °C. All measurements were carried out at a flow rate of 0.25 mL/min using a detector wavelength of 280 nm. The mass spectrometric data were collected in full-scan mode, from *m/z* 200 to 1000.

Efficiency of Piceatannol Extraction

The efficiency of piceatannol extraction was measured as follows. Five grams of ground passion fruit seeds were extracted three times by sonication (Shimadzu, Kyoto, Japan) at room temperature for 30 min using 50 mL of different concentrations of acetone, ethanol, methanol dichrolomethane, and ethylacetate. The supernatant was collected and the precipitate was sonicated using fresh solvent and collected. The process was repeated three times. The supernatant was measured for piceatannol content via HPLC, as described previously.

To examine further the efficiency of the extraction, the precipitate obtained after extraction with ethanol was extracted again with acetone. Each extraction was performed using the method described above. The supernatant was accumulated and piceatannol content was determined by HPLC.

Measurement of Melanin Content and Soluble Collagen Content

Sample Preparation

Passion fruit seeds were extracted as described above (for polyphenol quantification).

Cells and Cell Culture

Melanin-producing MNT-1 human melanoma cells (a gift from Dr. V. J. Hearing, Laboratory of Cell Biology, National Cancer Institute, National Institutes of Health, Bethesda, MD) were cultured in DMEM containing 10% FBS, 4 mM glutamine, 100 units/mL penicillin, and 100 μg/mL streptomycin.

SF-TY human dermal fibroblast cells (Health Science Research Resources Bank, Tokyo, Japan) were cultured in Medium 106S supplemented with 5% LSGS, 10 μg/mL gentamicin, and 0.25 μg/mL amphotericin B.

Melanoma and fibroblast cells were incubated at 37 °C under 5% CO_2 and 95% air.

Measurement of Melanin Content

MNT-1 cells were seeded at a density of 7.0×10^4 cells/well in 12-well culture plates and cultured for 24 h. Subsequently, the medium was replaced with fresh DMEM containing various concentrations of passion fruit seed extract or kojic acid [5-hydroxy-2-(hydroxymethyl)-4-pyrone], which was used as a positive control. After 72 h of culture, cells were washed with PBS and trypsinized with 0.25% trypsin containing 0.02% EDTA. The number of cells harvested was counted. To measure the melanin produced in these cells, harvested cells were washed twice with PBS and centrifuged at $300 \times g$ for 5 min at 4 °C, to obtain cell pellets. The cell pellets were dissolved in 500 μL of 1 N NaOH and quantified for melanin content using spectrophotometry at a wavelength of 415 nm. The concentration of the melanin produced was calculated based on standard curves obtained using synthetic melanin dissolved in 1 N NaOH.

Measurement of Soluble Collagen Content

SF-TY cells were seeded at a density of 1.7×10^4 cells/well in 48-well culture plates and cultured for 24 h. Subsequently, the medium was replaced with fresh medium containing various concentrations of passion fruit seed extract or ascorbate, which was used as a positive control. After 72 h of culture, the medium was collected and assayed for soluble collagen content using the Sircol™ Soluble Collagen Assay Kit, according to the manufacturer's instructions. Briefly, supernatants of SF-TY cells were centrifuged at $12{,}000 \times g$ for 4 min and 100 μL of each supernatant was mixed with 1 mL of Sircol dye and shaken for 30 min. The aliquot was then centrifuged at $12000 \times g$ for 10 min to pellet the collagen–dye complex. After decanting the suspension, droplets were dissolved in 0.75 mL of Sircol alkali reagent. The concentration of collagen was measured using spectrophotometry at a wavelength of 540 nm, using soluble collagen to obtain a standard curve. The value of the blank (i.e., the medium alone) was subtracted from each sample, to remove the contribution of the collagen contained in FBS. Adherent cells were washed twice with PBS and harvested using trypsin, for cell counting.

The melanin produced and the total soluble collagen were expressed as a concentration ratio relative to control cells that were not treated with the extracts or positive control.

Statistical Analyses

Data represent the mean ± standard deviation (SD) for the indicated number of experiments. The determination of the significance of the difference between the test sample and the corresponding control was performed using the paired t test, where $*$, $**$, $***$ represented $p < 0.1$, $p < 0.05$, and $p < 0.01$, respectively.

Results and Discussion

Determination of Polyphenol Content in Passion Fruit

Reports on the polyphenol content of passion fruit is limited, so we focused on the content and the kind of polyphenols contained in passion fruit. Total polyphenol content of the rind, pulp, and seeds contained 4, 0.2, and 33%, respectively. The seeds contained a much larger amount of polyphenols compared with the rind or pulp extract. The calculated polyphenol content in each part of the raw fruit was 0.22, 0.02, and 1.8%, respectively. The high polyphenol content contained in the seed showed high antioxidant activity, as assessed using superoxide dismutase (SOD) activity, was especially high in passion fruit seeds compared with the rind and pulp (data not shown). A good correlation between total phenol content and antioxidant capacity of tropical fruits were indicated (*21*), and similar results were obtained in this study. The phytochemical compounds of passion fruit juice have been studied (*7–9*), although there are not many on seeds. However, it is reported that passiflin, a novel protein found in passion fruit seed, acts as an antifungal protein (*22*) and that the insoluble fiber from passion fruit seed acts as a potential hypocholesterolemic ingredient (*23*). It is shown from our study that passion fruit seed contains a large amount of polyphenols which is high in antioxidant activity.

As the amount of polyphenols was high in passion fruit seeds compared with the rind and pulp, the polyphenol contained in the seeds was determined using chromatographic measurements. The specific peak 1, as depicted in Figure 1, corresponded to piceatannol (Figure 2A), which was present in large amounts in passion fruit seeds. The retention time and the UV spectrograph coincided with those of standard piceatannol, with piceatannol and peak 1 showing retention times of 19.01 and 19.13 min and UV (λ max) of 236, 303, and 324 nm and 238, 303, and 325 nm, respectively. The LC/MS data of passion fruit seed extracts also agreed with the presence of piceatannol, for which the $[M-H]^-$ peak was at *m/z* 243 (molecular weight of piceatannol, 244). Resveratrol (Figure 2B) was also detected and is represented in peak 2. The retention time and the UV spectrograph coincided with those of standard resveratrol, with resveratrol and peak 2 showing retention times of 23.43 and 23.61 min and UV (λ max) of 237, 306, and 319 nm and 236, 306, and 319 nm, respectively. The LC/MS data of passion fruit seed extracts also agreed with the presence of resveratrol, for which the $[M-H]^-$ peak was at *m/z* 227 (molecular weight of resveratrol, 228). There are reports indicating that resveratrol-containing plants may also contain piceatannol (*24*). Passion fruit seed contained both piceatannol and resveratrol. The LC/MS data of passion fruit seeds showed an $[M-H]^-$ peak at *m/z* 485, which suggests the presence of a dimer of piceatannol, the molecular weight of which is 486. Scirpusin B (Figure 2C), a dimer of piceatannol, was detected in passion fruit seeds and peak 3 in Figure 1 was determined as being scirpusin B in our study (*25*).

As shown in Figure 1, passion fruit seeds contained a high amount of piceatannol, which appeared as a large peak. To obtain piceatannol efficiently, extraction was performed using solvents such as ethanol, acetone, dichrolomethane, and ethylacetate. Figure 3 shows the amount of piceatannol extracted using each solvent. Extraction with ethanol and acetone was effective, especially at the concentrations of 80% and 70%, respectively. Furthermore, ethanol extraction followed by acetone extraction yielded a greater amount of piceatannol (343 mg/100 g freeze-dried seeds of piceatannol (Figure 4). Piceatannol was not detected after extraction with dichrolomethane and ethylacetate.

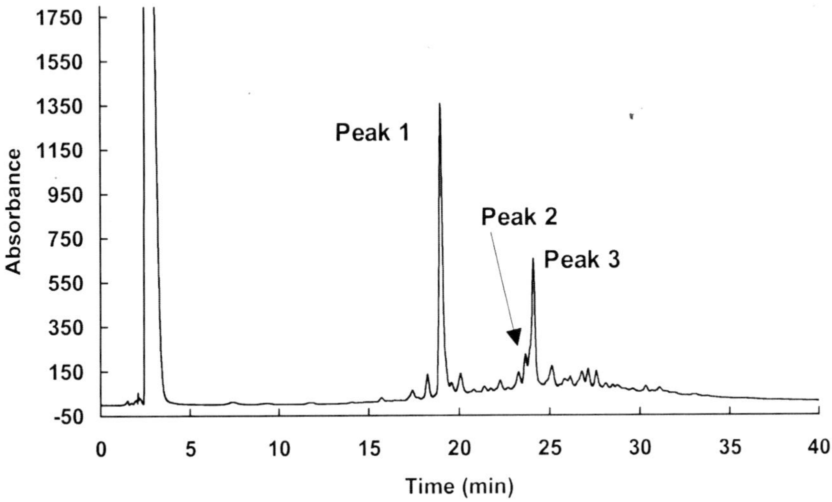

Figure 1. Determination of the concentration of piceatannol and resveratrol in passion fruit seeds using chromatographic measurements. Peak 1, peak 2, and peak 3 represent piceatannol, resveratrol, and scirpusin B, respectively. (Reproduced with permission from reference (26). Copyright 2010 ACS Publications).

Piceatannol is a tetrahydric polyphenol from the group of stilbene compounds, which the group includes resveratrol, pterostilbene and such. Resveratrol is one of the well-known polyphenols contained in grapes and peanuts (*11*, *27*) and reportedly contributes to skin photoprotection and antioxidation (*12*, *28*), among other functions. Reports on piceatannol are remarkably increasing and the content in plants, especially in grapes, is one of the focused studies recently. Moreover, piceatannol is metabolized from resveratrol via hydration by CYP1B1, which is involved in cancer prevention (*14*). There are many studies on the functions of piceatannol, which include vasorelaxing effects (*25*) and inhibition of melanin synthesis (*29*). Piceatannol is found in a limited number of plants, in limited amounts: 0.15% in dry fabaceae (*Caragana tibetica*) (*30*), 138–422 ng/g dry sample of highbush blueberry (*Vaccinium corymbosum*), and deerberry (*Vaccinium stamineum*) (*31*), and 0.052 μg/g fresh weight of grapes (*V. vinifera* L., cv. Cabernet Sauvignon) (*11*). Our results showed that passion fruit contains a very high content of piceatannol in a natural state, several times larger than that found in fabaccac (described above). Here, we report that piceatannol is present in passion fruit seeds and that the content of piceatannol in these seeds is much greater than that observed in any other plant. Piceatannol was not detected in the rind or pulp of passion fruit.

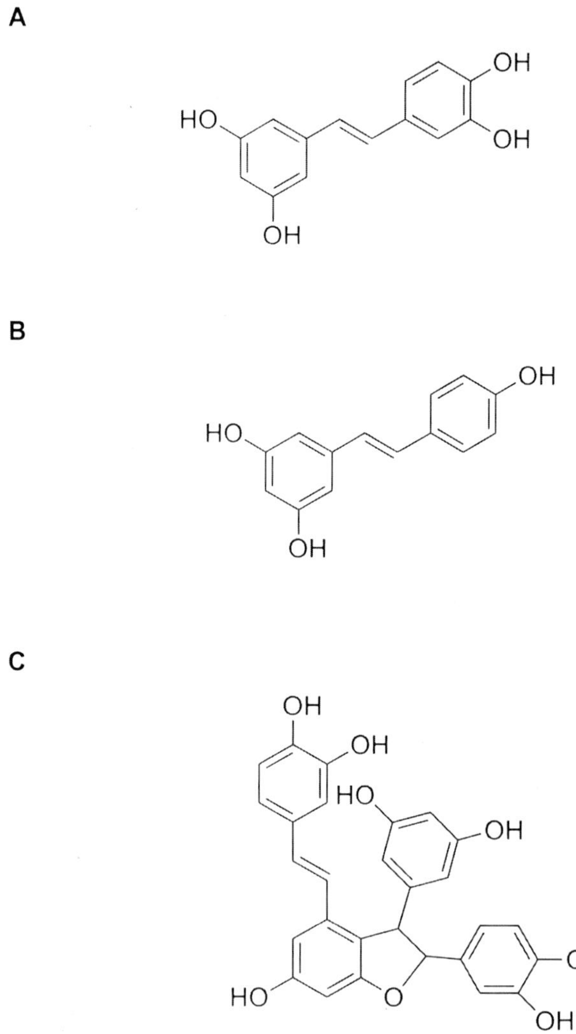

Figure 2. Chemical formulae of (A) piceatannol (3,4,3',5'-tetrahydroxy-trans-stilbene), (B) resveratrol (3,5,4'-trihydroxy-trans-stilbene), and (C) scirpusin B.

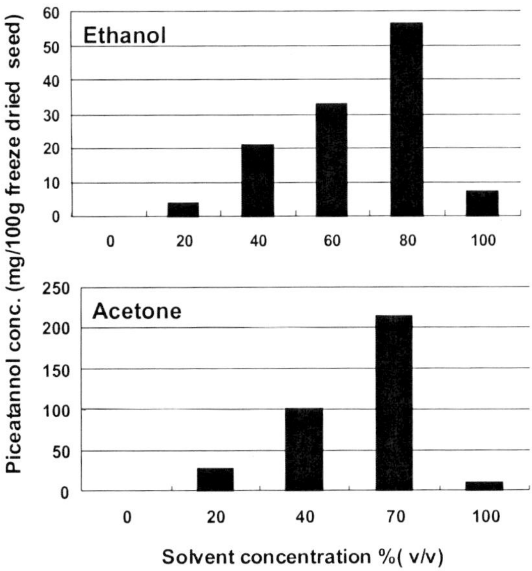

Figure 3. Extraction of piceatannol using ethanol or acetone.

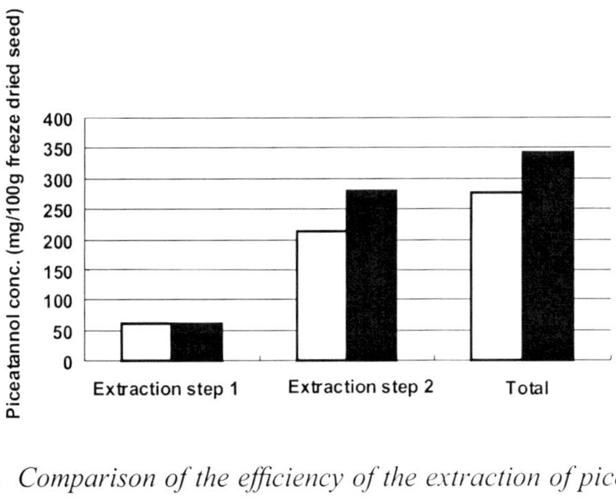

Figure 4. Comparison of the efficiency of the extraction of piceatannol. Individual extraction with ethanol (step 1) and acetone (step 2) is indicated by open bars, and ethanol extraction (step 1) following acetone extraction (step 2) is indicated by closed bars.

In Vitro Anti-Skin-Aging Effects of Passion Fruit Seeds

Polyphenols are thought to be natural antioxidants; polyphenols such as tea catechins (*32*) and proanthocyanidin (*16*) have been reported as having antioxidant activity in dermal cells. Exposure to UV rays, stress, medication, and many other environmental conditions lead to the production of reactive oxygen species (ROS) to an excessive degree, even though ROS are biogenic substances. The high antioxidant activity in passion fruit seeds led us to predict positive effects for the former on dermal cells, as reported previously.

Inhibition of melanin synthesis was measured in melanin-producing human melanoma cells. As shown in Figure 5, a significant decrease in melanin synthesis was observed after the application of passion fruit seed extracts to the melanoma cell culture, at 20 µg/mL.

Total soluble collagen was quantified in the culture medium of human dermal fibroblast cells (Figure 5). Soluble collagen synthesis increased significantly at a concentration of extract of 200 µg/mL. None of the samples exhibited inhibition of cell growth at the concentrations examined (data not shown).

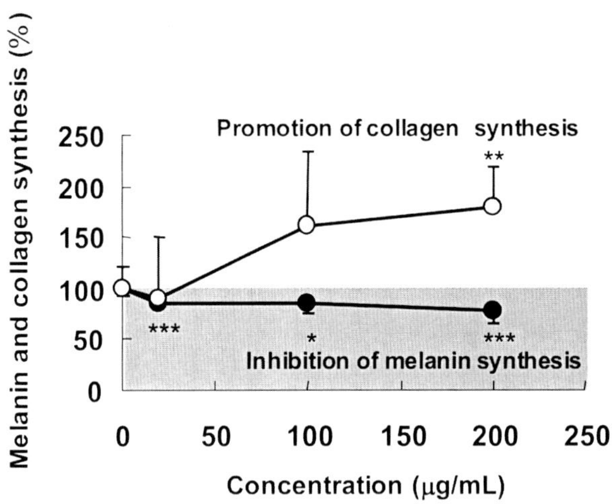

Figure 5. Melanin and collagen synthesis associated with the application of extracts of passion fruit seeds. Data are expressed as means ± SD (n = 3). (Data are from reference (26)).

Passion fruit seeds yielded positive effects on dermal cells, though the rind and pulp did not show any effects (data not shown). Luteolin, the polyphenol contained in the rind of passion fruit (*33*), inhibits melanogenesis (*34*); however, the results of our experiments using rind extracts did not show this inhibitory effect. *Schinus terebinthifoluis* Raddi extract combined with the linoleic acid fraction isolated from *Passiflora edulis* oil decreased melanin synthesis in B16 cells (*35*), which passion fruit seed may contribute to skin health not only by its high polyphenol content, but also by other compounds included in the seed. The oil content in the seed is approximately 23%, of which 72% is linoleic acid (*36*).

To evaluate the contribution of piceatannol to the effects of passion fruit seed extracts on the inhibition of melanin and promotion of collagen synthesis, piceatannol from the seed extracts and a commercially available piceatannol reagent were prepared to the same concentration, to compare their activities. As shown in Figure 6, piceatannol from seed extracts yielded a stronger effect regarding the inhibition of melanin synthesis, although the piceatannol reagent also yielded a strong inhibition of melanin synthesis. Both piceatannol samples decreased melanin synthesis significantly, at 4.5 μM. Soluble collagen synthesis was increased after treatment with both piceatannol samples; however, a significant difference between the two samples was not identified.

Piceatannol is often compared with resveratrol because of their similar chemical structures. We compared the melanin-synthesis-inhibitory and collagen-synthesis-promoting effects of piceatannol and resveratrol, which are both contained in passion fruit seed extracts. As shown in Figure 7, both compounds inhibited melanin synthesis, although piceatannol yielded a higher melanin inhibitory effect compared with resveratrol. Piceatannol led to a significant decrease in melanin synthesis at 5 μM, whereas resveratrol led to a significant decrease in melanin synthesis at 20 μM. Moreover, both piceatannol and resveratrol increased soluble collagen: piceatannol increased soluble collagen significantly at 5 μM, whereas resveratrol increased soluble collagen significantly at 10 μM.

Piceatannol and resveratrol both led to inhibition of melanin synthesis and production of soluble collagen; however, piceatannol yielded stronger effects in both dermal cell types. The effect of piceatannol on dermal cells has been reported. The melanin-synthesis inhibitory activities of piceatannol and resveratrol were studied in melanoma cells, showing that the melanin-synthesis-inhibitory activity of piceatannol was higher than that of resveratrol because of the higher antioxidant activity of piceatannol (*29*). Similar results were obtained in this study. Several mechanisms for the melanin-synthesis-inhibitory activity is reported such as inhibition of tyrosinase activity (*15*, *16*) and inhibition of cytokines such as endothelin-1 (ET-1) (*37*) or α-melanocyte stimulating hormone (α-MSH) (*38*). The antioxidant activity of the polyphenols contained in passion fruit seed extract may have contributed in such enzymatic reaction to give melanin-synthesis-inhibitory activity. One of the major mechanisms of collagen-synthesis activity is the inhibition of the degradation of collagen via inhibition of matrix metalloproteinases (MMPs), and it is an effective approach to the maintenance of collagen in the dermis. Piceatannol is an inhibitor of the JAK1/STAT-1 pathway, which induces the expression of the *MMP-1* gene in cultured human dermal fibroblasts (*14*). Moreover, catechins, which are well-known polyphenols, also inhibit the UV-induced expression of MMPs (*18*). These reports led us to speculate that the PF-S-induced increase in the levels of collagen may be the result of the inhibition of MMPs or of their activity by the polyphenols present in PF-S; however, additional experiments are required to elucidate these issues.

The larger number of hydroxyl groups present in piceatannol may contribute to its stronger effect compared with resveratrol; in addition, the catechol structure may yield a high antioxidant effect. 1,1-diphenyl-2-picrylhydrazyl radical-scavenging assays of piceatannol, scirpusin B, and resveratrol have been reported. Radical-scavenging activity was high in the order of scirpusin B, piceatannol, and resveratrol (*25*). Also, another study shows that the antioxidant activity of the dimer of resveratrol is stronger than resveratrol (*39*).This led us to speculate that the stronger effect of the passion fruit seed extract compared with the piceatannol and resveratrol reagents is due to the presence of resveratrol and scirpusin B in the extract. The results described above suggest that the passion fruit seed extract may have a beneficial effect on dermal cells because of its high antioxidant activity. However, further studies are needed to identify the precise mechanism underlying these effects.

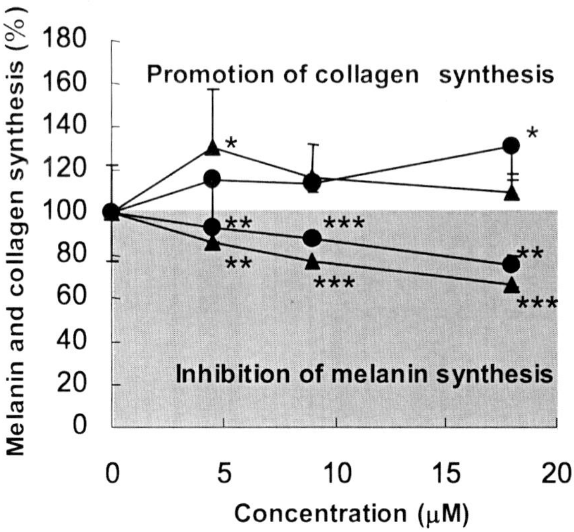

Figure 6. Comparison of the effect of the passion fruit seed extract (▲) and piceatannol reagent (○) on melanin and collagen synthesis. Data are expressed as means ± SD (n = 3). (Data are from reference (26)).

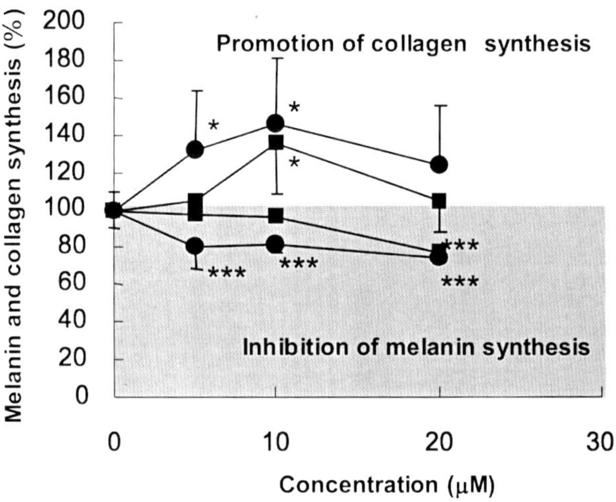

Figure 7. Comparison of the effect of piceatannol (○) and resveratrol (■) reagents on melanin and collagen synthesis. Data are expressed as means ± SD (n = 3). (Data are from reference (26)).

Conclusion

Piceatannol is contained in passion fruit seeds at very high levels and exerted positive effects on dermal cells: inhibition of melanogenesis and promotion of the synthesis of collagen. Moreover, the derivatives of piceatannol, resveratrol, and scirpusin B contained in passion fruit seed may contribute to these effects. It is strongly suggested that the high antioxidant activity of the extract underlies these effects, as the synthesis of melanin and the degradation of collagen are accompanied by the production of ROS. It is possible to suggest that oral and topical application of passion fruit seed extracts may contribute to a decrease in skin damage, which may cause skin melanogenesis and wrinkles, and that passion fruit seeds may be a novel promising material to prevent skin abnormalities.

References

1. Gorinstein, S.; Zemser, M.; Haruenkit, R.; Chuthakorn, R.; Grauer, F.; Martin-Belloso, O.; Trakhtenberg, S. *J. Nutr. Biochem.* **1999**, *10*, 367–371.
2. Mertens-Talcott, S. U.; Rios, J.; Jilma-Stohlawetz, P.; Pacheco-Palencia, L. A.; Meibohm, B.; Talcott, S. T.; Derendorf, H. *J. Agric. Food Chem.* **2008**, *56*, 7796–7802.
3. Kondo, M.; Zhang, L.; Ji, H.; Kou, Y.; Ou, B. *J. Agric. Food Chem.* **2009**, *57*, 8788–8792.
4. Movafegh, A.; Alizadeh, R.; Hajimohamadi, F.; Esfehani, F.; Nejatfar, M. *Anesth. Analg.* **2008**, *106*, 1728–1732.
5. Montanher, A. B.; Zucolotto, S. M.; Schenkel, E. P.; Fröde, T. S. *J Ethnopharmacol.* **2007**, *10*, 281–288.
6. Dhawan, K.; Sharma, A. *Fitoterapia* **2002**, *73*, 397–399.
7. Talcott, S. T.; Percival, S. S.; Pittet-Moore, J.; Celoria, C. *J. Agric. Food Chem.* **2003**, *51*, 935–941.
8. Mercadante, A. Z.; Britton, G.; Rodriguez, D. B. *J. Agric. Food Chem.* **1998**, *46*, 4102–4106.
9. Suntornsuk, L.; Gritsanapun, W.; Nilkamhank, S.; Paochom, A. *J. Pharm. Biomed. Anal.* **2002**, *28*, 849–855.
10. Goldberg, D. M.; Hahn, S. E.; Parkes, J. G. *Clin. Chim. Acta.* **1995**, *237*, 155–187.
11. Bavaresco, L.; Fregoni, M.; Trecisan, M.; Mattivif, F.; Vrhovsk, U.; Falchetti, R. *Vitis.* **2002**, *41*, 133–136.
12. Vergara, C.; von Baer, D.; Mardones, C.; Wilkens, A.; Wernekinck, K.; Damm, A.; Macke, S.; Gorene, T.; Winterhalter, P. *J. Agric. Food Chem.* **2012**, *60*, 929–933.
13. Hung, L. M.; Su, M. J.; Chu, W. K.; Chiao, C. W.; Chan, W. F.; Chen, J. K. *Br. J. Pharmacol.* **2002**, *135*, 1627–1633.
14. Potter, G. A.; Patterson, L. H.; Wanogho, E.; Perry, P. J.; Butler, P. C.; Ijaz, T.; Ruparelia, K. C.; Lamb, J. H.; Farmer, P. B.; Stanley, L. A.; Burke, M. D. *Br. J. Cancer.* **2002**, *86*, 774–778.
15. Roh, J. S.; Han, J. Y.; Kim, J. H.; Hwang, J. K. *Biol. Pharm. Bull.* **2004**, *27*, 1976–1978.

16. Yamakoshi, J.; Otsuka, F.; Sano, A.; Tokutake, S.; Saito, M.; Kikuchi, M.; Kubota, Y. *Pigment Cell Res.* **2003**, *16*, 629–638.
17. Koya-Miyata, S.; Okamoto, I.; Ushio, S.; Iwaki, K.; Ikeda, M.; Kurimoto, M. *Biosci. Biotechnol. Biochem.* **2004**, *68*, 767–773.
18. Vayalil, P. K.; Mittal, A.; Hara, Y.; Elmets, C. A.; Katiyar, S. K. *J. Invest. Dermatol.* **2004**, *122*, 1480–1487.
19. Lim, H.; Kim, H. P. *Planta Med.* **2007**, *73*, 1267–1274.
20. Sanoner, P.; Guyot, S.; Marnet, N.; Molle, D.; Drilleau, J. P. *J. Agric. Food Chem.* **1999**, *47*, 4847–4853.
21. Martinez, R.; Torres, P.; Menesses, M. A.; Figueroa, J. G.; Perez-Alvarez, J. A.; Viuda-Martos, M. *Food Chem.* **2012**, *135*, 1520–1526.
22. Lam, S. K.; Ng, T. B. *Phytomedicine* **2009**, *16*, 172–180.
23. Chau, C. F.; Huang, Y. L. *Mol. Nutr. Food Res.* **2005**, *49*, 786–790.
24. Lin, L. L.; Lien, C. Y.; Cheng, Y. C.; Ku, K. L. *J. Chromatogr. B Anal. Technol. Biomed. Life Sci.* **2007**, *853*, 175–182.
25. Sano, S.; Sugiyama, K.; Ito, T.; Katano, Y.; Ishihata, A. *J. Agric. Food Chem.* **2011**, *59*, 6209–6213.
26. Matsui, Y.; Sugiyama, K.; Kamei, M.; Takahashi, T.; Suzuki, T.; Katagata, Y.; Ito, T. *J. Agric. Food Chem.* **2010**, *58*, 11112–11118.
27. Sanders, T. H.; McMichael, R. W., Jr.; Hendrix, K. W. *J. Agric. Food Chem.* **2000**, *48*, 1243–1246.
28. Nichols, J. A.; Katiyar, S. K. *Arch. Dermatol Res.* **2010**, *302*, 71–83.
29. Yokozawa, T.; Kim, Y. J. *Biol. Pharm. Bull.* **2007**, *30*, 2007–2011.
30. Xiang, T.; Uno, T.; Ogino, F.; Ai, C.; Duo, J.; Sankawa, U. *Chem. Pharm. Bull.* **2005**, *53*, 1204–1206.
31. Rimando, A. M.; Kalt, W.; Magee, J. B.; Dewey, J.; Ballington, J. R. *J. Agric. Food Chem.* **2004**, *52*, 4713–4719.
32. Jeon, S. E.; Choi-Kwon, S.; Park, K. A.; Lee, H. J.; Park, M. S.; Lee, J. H.; Kwon, S. B.; Park, K. C. *Photodermatol. Photoimmunol. Photomed.* **2003**, *19*, 235–241.
33. Ichimura, T.; Yamanaka, A.; Ichiba, T.; Toyokawa, T.; Kamada, Y.; Tamamura, T.; Maruyama, S. *Biosci. Biotechnol. Biochem.* **2006**, *70*, 718–721.
34. Choi, M. Y.; Song, H. S.; Hur, H. S.; Sim, S. S. *Arch. Pharm. Res.* **2008**, *31*, 1166–1171.
35. Jorge, A. T.; Arroteia, K. F.; Santos, I. A.; Andres, E.; Medina, S. P.; Ferrari, C. R.; Lourenco, C. B.; Biaggio, R. M.; Moreira, P. L. *Int. J. Cosmet. Sci.* **2012**, *34*, 435–440.
36. Liu, S.; Yang, F.; Li, J.; Zhang, C.; Ji, H.; Hong, P. *Int. J. Food Sci. Nutr.* **2008**, *59*, 706–715.
37. Imokawa, G.; Yada, Y.; Miyagishi, M. *J. Biol. Chem.* **1992**, *267*, 24675–24680.
38. Thody, A. J.; Graham, A. *Pigment Cell Res.* **1998**, *11*, 265–274.
39. Kim, H. J.; Chang, E. L.; Cho, S. H.; Chung, S. K.; Park, H. D.; Choi, S. W. *Biosci. Biotechnol Biochem.* **2002**, *66*, 1990–1993.

Chapter 13

Bioactive Compounds in *Moringa oleifera*: Isolation, Structure Elucidation, and Their Antiproliferative Properties

Bishambar Dayal,[*,1,2] Vineela Reddy Yannamreddy,[2] Ritesh Amin,[2] Michael A. Lea,[2] and Athula B. Attygalle[3]

[1]Department of Medicine, UMDNJ-New Jersey Medical School, Newark, New Jersey 07103
[2]Department of Biochemistry and Molecular Biology, UMDNJ-New Jersey Medical School, Newark, New Jersey 07103
[3]Department of Chemistry, Chemical Biology and Biomedical Engineering, Stevens Institute of Technology, Hoboken, New Jersey 07030
[*]E-mail: dayalbi@umdnj.edu.

A systematic study for the isolation and structure elucidation using thin-layer chromatography (TLC)/electrospray ionization mass spectrometry (ESI-MS) of microwaved methanolic extract of different sections of the tropical fruit *Moringa oleifera* is described. Different tissues of *Moringa oleifera* such as seed coat (inner skin skeleton), pulp and seeds when subjected to electrospray ionization mass spectrometry provided 4-(±-L-rhamnopyranosyloxy) benzyl glucosinolate and its corresponding three mono acetyl isomers. Benzyl glucosinolate at *m/z* 408, its parahydroxy derivative and its higher homologue are also profiled. Biological active myrosinase hydrolyzed products, 1-iso thiocyanate-(5-methylsulfinylpentane), and 1-isothiocyanato-3-methylsulfonyl-propane presumably by the hydrolysis catalyzed by myrosinase were also characterized. The identification of O-ethyl-4-[α-L-rhamnosyloxy benzyl]-carbamate from *M. oleifera* pulp provided m/z 341 is also described. Collision-induced dissociation mass spectra of anions present in *Moringa oleifera* exhibited product ion at *m/z* 97, characteristic of sulfate moiety (HSO_4^-) in the molecular structure of glucosinolates. The proliferation inhibitory effects

© 2013 American Chemical Society

of *Moringa oleifera* seed, seed coat, pulp (inner skin skeleton) and skin on human colon cancer HT-29 cells and Caco-2 cells are investigated in these studies.

Introduction

Moringa oleifera Lamarck (Moringaceae), also known as drumstick tree, horseradish tree, is a fast growing kelor tree widely distributed in India, Bangladesh and Afghanistan (*1*). *Moringa oleifera* found in tropical and subtropical areas, including sub- or Himalayan regions of India, South Africa, Arabia, Philippines, Cambodia, South America, Pacific Islands, and Caribbean Islands is known for its medicinal qualities for thousands of years (*1*, *2*). *M. oleifera, parts* such as young leaves, flowers, and green pods, have been used for their nutritional value (*3*). In addition, it has been widely used in the Unani and Sidha medicine and also as a traditional herbal remedy for the treatment of a variety of disorders such as skin diseases, respiratory sickness, ear and dental infections, hypertension, diabetes, anemia, and cancer (*4–6*). Pharmacological properties include antihypertensive, diuretic, cholesterol lowering, antispasmodic, antiulcer, hepatoprotective, antibacterial, antifungal, antitumor/ anticancer, antioxidant, and antihelmintic activities (*3–15*). Furthermore, the seed of the plant is an effective natural coagulant and can be used to purify water (*16*) and as a possible source of biodiesel (*17*). Some glycosidic constituents glucosinolates, known to contribute to human health promotion due to their enzymatically induced hydrolyzed products (*18–21*). When the cell wall is disrupted, it releases the enzyme myrosinase, which hydrolyzes glucosinolates to produce primarily isothiocyanates, allylisothiocyanate and other breakdown products (*20–25*). Such compounds have been isolated and identified (e.g., niazirin, niazimicin, and niazicin A) and reported to mediate antitumor or anti-inflammatory activities (*7*, *8*, *26*, *27*). Present studies describe the isolation and characterization of glucosinolates, their hydrolyzed products, from the seeds and seed coat of *Moringa oleifera*. The identification of O-ethyl-4-[α-L-rhamnosyloxy benzyl]-carbamate from *M. oleifera* pulp is also described. The data presented in these studies suggest potential use of *Moringa oleifera* seeds and seed coat as antioxidants and antiproliferative agents.

Experimental

Plant Materials

Drumsticks (*Moringa oleifera*) were obtained from a local market (*Edison, New Jersey, USA)* and dissected into four different sections (Figure 1), namely, skin, pulp, seed coat, and seeds.

Figure 1. Fresh Moringa oleifera, dissected sections and their seeds.

Microwave-Assisted Extraction

The individual sections were treated with methanol and microwaved for efficient extraction of the biologically active compounds as described previously (Figure 2) (*28–30*). Briefly, 500 mg seeds were extracted with 2 ml of methanol and microwaved for a total of 42 seconds in a domestic microwave oven with 6 seconds intervals. Caution is to be excercised because the low boiling solvent methanol evaporates very quickly and the higher temperatures result in significant degradation of glucosionolates. This procedure always resulted in 90% recovery of the bioactive compounds. The methanol solvent was evaporated under nitrogen atmosphere and the resulting isolated bioactives were obtained in (90% yield).

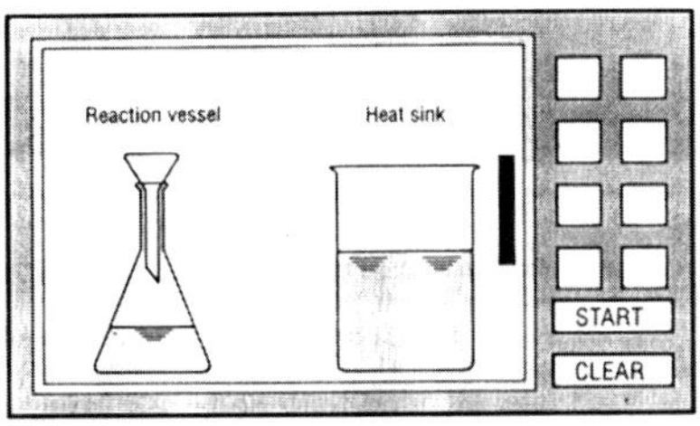

Figure 2. Domestic microwave equipment used for the extraction of bioactives from skin, pulp, seed coat, and seed using methanol.

Using on/off heating cycle, it is convenient to fine tune the energy absorbed by the reaction mixture by placing a beaker of water or *N, N*-dimethyl formamide (DMF) near the reaction vessel. This extra polar liquid acts as a heat sink by absorbing a portion of the microwave energy (*28–36*).

Extraction of Intact Glucosinolates from Seeds by Deactivating the Enzyme Myrosinase

Since in intact cells, glucosinolates remain separate entity from myrosinase which hydrolyzes them, therefore it is desirable to adopt a extraction methodology to isolate them without their breakdown products.

Glucosinolates were extracted from seeds based on a previously described procedure (*25*). Seeds (0.1 g) of a *Moringa oleifera* were placed in a closed vial submerged in boiling water for 15 min to deactivate the enzyme myrosinase. The seeds were then pulverized and extracted three times with boiling water–methanol (19:1). After each hot water–methanol extraction, samples were allowed to cool and were centrifuged at 11,000 rpm for 2–3 min. The supernatant layers were separated, combined and passed through a DSC-18 plug (Supelco / Sigma Aldrich, St. Louis Mo USA). The eluent was concentrated under reduced pressure to yield a crude mixture of glucosinolates. This concentrated extract was taken up in 0.5 ml of water and centrifuged at 11,000 rpm for 2 min to give the final seed extract as the supernatant.

Thin-Layer Chromatography

Chromatographic Materials

All TLC isolated products were detected as one spot when subjected to thin-layer chromatographic examination (precoated 0.25-mm silica plates from Analtech, Newark, DE) (*35*).

The compounds were resolved and analyzed by analytical and preparatory TLC and subjected to electrospray-ionization mass spectrometry. Microwaved methanolic extract (30 mg) after evaporating under nitrogen was further sequentially extracted with hexane, and chloroform.The chloroform and hexane extracts from the seed coat and seed cover were then separated in the following TLC solvent system (chloroform: methanol: acetic acid 18:4:0.5 v/v/v). The hexane extract provided low-molecular weight compounds and chloroform extract provided a mixture of intact glucosinolates, their acetyl derivatives, isothiocyanates, rhamnose carbamate as revealed by negative ion electrospray ionization mass spectrometry. Further MS/MS (MS2 scans) were performed for peaks at m/z 341 and 570 which demonstrated structural assignment for 4-(+α L-rhamnopyranosyloxy) benzyl glucosinolate (Figure3) and rhamnose carbamate (Figure 9).

Determination of Total Phenolics and DPPH Radical Scavenging Activity

Assays were performed with methanolic extracts of Moringa oleifera seeds, seed coat, inner skin skeleton and pulp. Phenolics were measured using the Folin-Ciocalteau reagent and expressed as mg chlorogenic acid equivalent per gram wet weight. Anti-oxidant activity was determined using a 2, 2 diphenyl-1-picrylhydrazyl assay and is expressed as mg trolox equivalent per gram wet weight as described previously (*15*, *35*).

Collision-Induced (CID) Mass Spectra

The Collision-induced (CID) mass spectra of glucosinolates were recorded on a Micromass Quattro I tandem mass spectrometer equipped with an electrospray ion source as described previously (*37*, *38*). All the dissected sections of *Moringa oleifera* were dissolved in acetonitrile–water–ammonia (9:1:0.00001, v/v/v) solution and were run at a flow rate of 6 μl min^{-1} as described previously (*37*,

38). The source temperature was held at 80 °C. The capillary voltage was held at 4000 V. The argon gas pressure in the collision cell was adjusted to attenuate precursor transmission by 30–50%. Accurate mass measurements were performed at high resolution (R ≈ 15000) on a Waters-Micromass Q-TOF API-US spectrometer equipped with a nanoelectrospray ion source. Signals were acquired in the W-mode operation. Ion series generated by water clusters charged by the bicarbonate anion were used as reference mass peaks. The source temperature was held at 80 °C. The capillary voltage was held at 3000 V. Pressure of argon gas in the collision cell was held at 5.1×10^{-5} Pa.

Cell Proliferation Assay

HT-29 and Caco-2 cells were obtained from American Type Culture Collection, Rockville, MD, USA, and were incubated at 37 °C in RPMI-1640 medium with 5% fetal calf serum.

A total of 1.5×10^6 cells were incubated with 10 mL RPMI-1640 medium with 5% fetal calf serum. The medium was changed after 24 hours and the cells were incubated with control or fractions of *Moringa oleifera* samples for 72 hours before harvesting (*39*, *40*). Inhibition of growth of Caco-2 human colon cancer cells after incubation for 72 hours with fractions from 200 μg/mL of inner skin skeleton/pulp (B6) , 200 μg/mL of seed coat (C6) and 200 μg/mL of seeds (D6) of *Moringa oleifera* is illustrated in Figure 9 (*39–41*). The cells were washed with phosphate-buffered saline and extracted with 0.5% NP40, 0.25 M NaCl, 5 mM EDTA and 50 mM Tris pH 8.0. The protein concentration of the extract was determined using the BCA Protein Assay Reagent from Pierce, Rockford, IL, USA.

Cell proliferation was also monitored using the WST-1 reagent according to the directions of the supplier, Roche Diagnostics, Mannheim, Germany. The difference in absorbance at 450 and 695 nm was recorded as described previously (*39–41*).

Results and Discussion

As part of our research program, we describe the isolation and structure elucidation of intact glucosinolates, their hydrolysis products, and their antiproliferative activities.

We have carried out microwave-heating extraction methodology for isolation of various *Moringa oleifera* fractions in our laboratory (*30–32*). This technique is simple, fast and highly efficacious. Recovering of glucosinolates and their hydrolyzed products were obtained in 80-90% yield. Microwave assisted extraction of Withanolides from roots of Ashwagandha and its comparison with conventional extraction method has been recently highlighted by Jyothi et al. (*33*). Furthermore, we have reported highly efficient microwave-induced organic synthesis of sterols and bile acid reactions such as hydrogenation, esterification, formylations, deactylation in high yield in a domestic microwave-oven (*28–35*). Bose et al. have cited microwave oven as the Bunsen burner of the 21st century (Chem. Eng. News Sept. 24, 2012, pp 32-34) (*36*).

The total antioxidant and phenolic activity of extracts of *Moringa oleifera* showed that the outer skin has the highest contribution of phenols and antioxidants among the different tissues. On the other hand, the seeds have the lowest phenolic content and antioxidant activity (*35*).

Identification of Glucosinolates by Collision-Induced Electrospray-Ionization Mass Spectrometry of *Moringa oleifera* Metabolites

Identification of glucosinolates and their hydrolysis products was accomplished by collision-induced electrospray-ionization mass spectrometry (*37*, *38*). The individual compounds were structurally characterized in the *-ve* ion mode. The following *m/z* values and fragment ions were obtained which were characteristic for these molecules.

4-(α-L-Rhamnopyranosyloxy) benzyl glucosinolate TLC fraction seed coat, R_f 0.23) exhibited molecular ion peaks at *m/z* 570 (100%), *m/z* 97 (90%, $HSO_4)^-$] *m/z* 96 (83%, $[SO_4]^-$) (Figure 4). The other characteristic fragment ion, *m/z* 195 was assigned to 1-thiohexose anion. Two minor peaks at *m/z* 259, 275 represented glucose 1-sulfate anion formed by the loss of phenyl isothiocyanates and glucose 1-thiosulfate anion respectively. Therefore, the ion at *m/z* 570 (100% intensity) from seed coat was assigned to 4-(±-L-rhamnopyranosyloxy) benzyl glucosinolate (Figure 3).

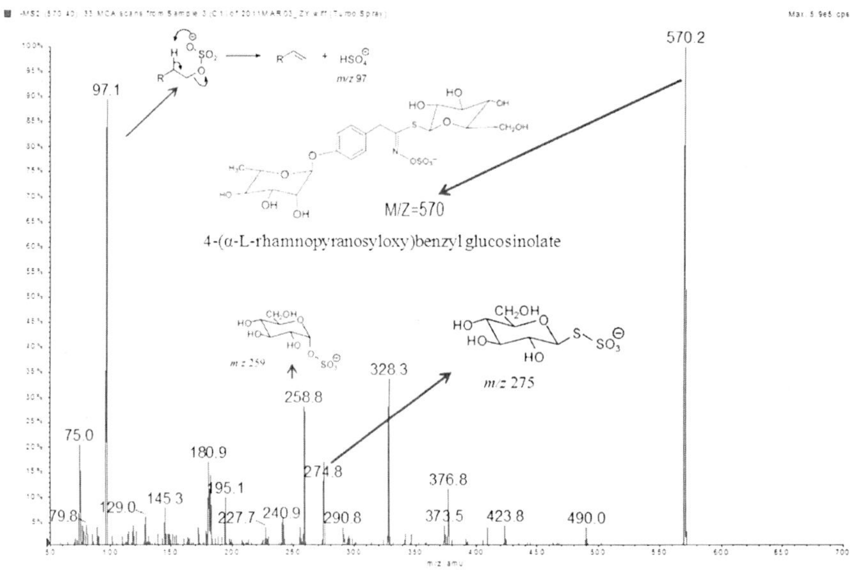

Figure 3. ESI-MS (-ve ion) of Moringa oleifera seed coat (C).

First organic synthesis of this major glucosinolate isolated from Moringa oleifera seeds and its structural characterization has been achieved by Gueyrard (*9*). Although MS fragmentation has not been described in detail but proton and carbon NMR spectroscopy data confirmed the stereochemistry assignment as 4-(α-L-rhamnopyranosyloxy) benzyl glucosinolate. Briefly, mass spectral data described in the (Ionspray® mode) by Gueyrard et al. provided *m/z* 570 (M-H$^+$). Proton and carbon NMR spectroscopy given below aided the structure and stereochemical assignment to this molecule as shown in (Figure 4).

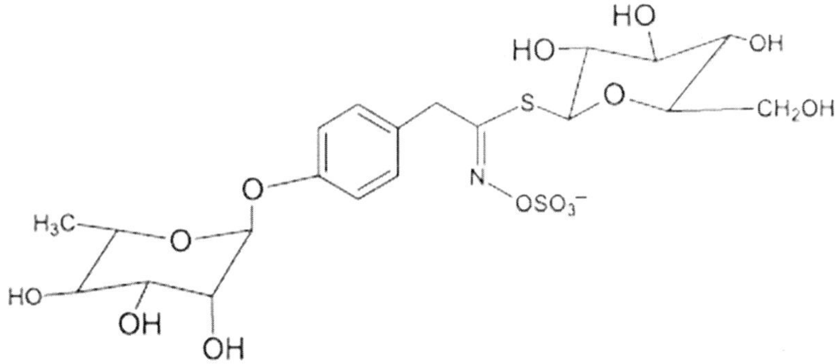

Figure 4. 4-(±-L-rhamnopyranosyloxy) benzyl glucosinolate.

^{1}H NMR (250 MHz), Figure 4: *δ*: 1.26 (d, 3H, H-6), 3.26 (m, 1H, H-5), 3.32–3.48 (m, 3H, H-2, H-3 and H-4) and 4.74 (m, 1H, H-1), are characteristic of rhamnose moiety and 3.55 (t, 1H, *J*=9.7, H-4′), 3.64–3.72 (m, 2H, H-6′), 3.83 (m, 1H, H-5′), 4.03 (dd, 1H, *J*=3.5, H-3′), 4.20 (dd, 1H, *J*=1.9, H-2′), and 5.58 (bs, 1H, H-1′), are assigned to glucose molecule and 4.13 (s, 2H, H-8) is benzylic methylene and 7.18 (d, 2H, *J*=8.5, H-11), 7.40 (d, 2H, H-10) are assigned to four protons in the aromatic ring as shown in the Figure 4 (*9*).

^{13}C NMR (62.89 MHz), Figure 4: 18.2 (C-6′), 39.0 (C-8), 61.8 (C-6), 70.2 (C-4), 70.9 (C-5′), 71.5 (C-2′), 71.6 (C-3′), 73.3 (C-2), 73.5 (C-4′), 78.5 (C-3), 82.9 (C-1), 83.4 (C-5), 99.6 (C-1′), 119.1 (C-11), 130.9 (C-9 and C-10), 156.2 (C-12), 164.2 (C-7) (*9*).

4-[(2′-*O*-acetyl-α-L-Rhamnopyranosyloxy)benzyl] glucosinolate (R_f 0.24, fraction 2 from seed coat/seed Figure 5). A close examination of the acetyl derivative of intact glucosinolate peak clearly shows *m/z* 612 in three isomeric forms with three different intensities (R_f, 0.24, Figure 5). Amaglo and Bennett et al. have reported the three isomeric structures of 4-[(2′-O-acetyl-α-L-rhamnopyranosyloxy) benzyl] glucosinolate in the *Moringa oleifera* root (*22*, *42*).

Isomer I: 4-[(2′-*O*-acetyl-α-L-rhamnopyranosyloxy) benzyl] glucosinolate, *m/z* 612 (15%, [M-H]$^-$), *m/z* 97 (75%, [SO_4H]$^-$), *m/z* 96 (50%, [SO_4]$^-$).

Isomer II: 4-[(2′-*O*-acetyl-α-L-rhamnopyranosyloxy) benzyl] glucosinolate, *m/z* 612 (16%, [M-H]$^-$), *m/z* 97 (72.7%, [SO_4H]$^-$), *m/z* 96 (50, [SO_4]$^-$).

Isomer III: 4-[(2′-*O*-acetyl-α-L-rhamnopyranosyloxy) benzyl] glucosinolate, *m/z* 612 (17%, [M-H]$^-$), *m/z* 97 (72.0%, [SO_4H]$^-$), *m/z* 96 (50%, [SO_4]$^-$).

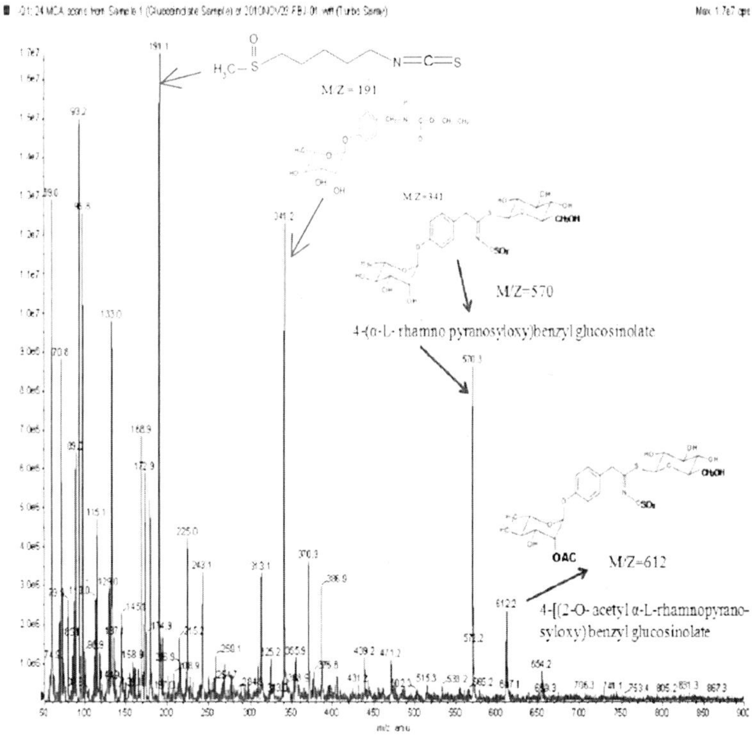

Figure 5. ESI-MS (-ve ion) of Moringa oleifera pulp (B).

The acetyl derivative of intact glucosinolate provided *m/z* 612 (Figure 5). To our knowledge, the intact acetyl derivative of 4-(±-L-rhamnopyranosyloxy) benzyl glucosinolate from inner skin skeleton or seed coat has not been reported in the literature. Leaves of the *Moringa oleifera* plant containing (4-[(2′-*O*-acetyl-α-L-rhamnopyranosyloxy) benzyl] glucosinolate and its corresponding three monoacetyl isomers (Figure 4) accompanied by various amino acids such as aspartic acid, phenylalanine, tryptophan, cysteine, methionine have been reported (*3*).

ESI-MS (-ve Ion) of Moringa oleifera Seed Coat Chloroform Fraction Designated as Glucotropaeolin

The peaks at *m/z* 408 and 425, as shown in Figure 6, have been rationalized for the formation of benzyl glucosinolate (glucotropaeolin) corresponding to *m/z* 408 with the following intensities: (18%, $[M-H]^-$), *m/z* 97 (100%, $[SO_4H]^-$), *m/z* 96 (90%, $[HSO_4]^-$). The peaks at *m/z* 425 and 439 have been rationalized as parahydroxy benzyl glucosinolate and its corresponding 14 units higher homologue as parahydroxy methyl benzyl glucosinolate. These compounds have been characterized and studied *in vitro* and *in vivo* for their medicinal properties (*20*, *23*, *43*).

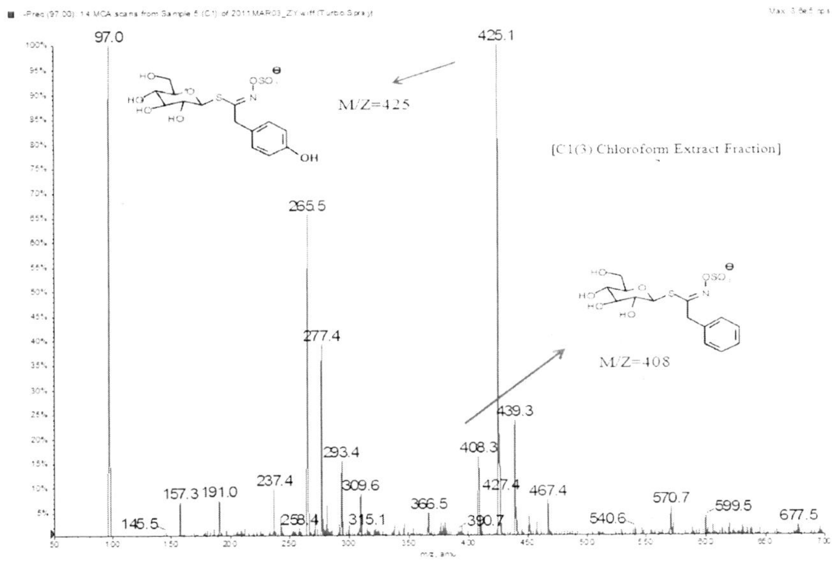

Figure 6. ESI-MS (-ve ion) of Moringa oleifera seed coat chloroform fraction.

The proposed mass spectral fragmentation and rearrangement to form glucose 1-thiosulfate anion at *m/z* 275 and 1-thiohexose anion at *m/z* 259 has been rationalized by Bialecki and coworkers in their collision-induced dissociation mass spectra of glucosinolate anions (*37*).

The three aforementioned isomeric forms of acetyl derivatives and their hydrolyzed products have several antimicrobial and anticarcinogenic properties ((*6*, *18*, *25*, *42–48*), Figures 7,8). We believe *Moringa oleifera* may be useful in the prevention and treatment of cancer and neurodegenerative diseases. The parent glucosinolates have little biological activity by themselves, but are converted to active products ((*42–48*), Figure 8) by the enzyme myrosinase in conjunction with other proteins.

Glucosinolates are sulfur-containing glycosides and are derived from *Brassica* vegetables, such as broccoli, cabbage, and watercress. These compounds are hydrolyzed enzymatically by the enzyme myrosinase after cell disruption to form isothiocyanates, nitriles, and other breakdown products ((*42–48*), Figures 7, 8).

Characterization of CH_3-SO-$(CH_2)_5$-NCS (compound 6, Figure 7): The peak at *m/z* 191 (100% intensity Figure 5) has been assigned to CH_3-SO-$(CH_2)_5$-NCS, a 14 units higher analog of the parent suforaphane (compound 3, Figure 7) and the hydrolyzed product of corresponding glucosinolate. In a recent and important study by Gurion et al. have elucidated that two natural isothiocyanates isolated from broccoli namely sulforaphane and erucin are potent inhibitors that inhibit bacterial quorum sensing and thus fight dangerous pathogens (*44*).

Furthermore, plant tissue damage is also possible, which allows the enzyme myrosinase to come in contact with glucosinolates, thus removing the glucose moiety from the glucosinolates (Figure 8). This results in an unstable species that rearranges to form isothiocyanates, nitriles, and other compounds (Figures 7,8).

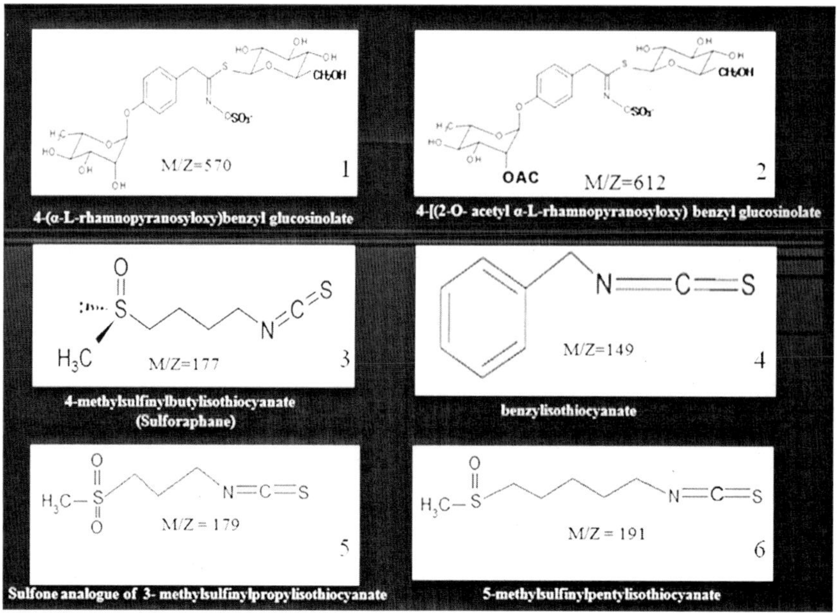

Figure 7. Structures of glucosinolates [1 and 2] and their enzymatically hydrolyzed isothiocyanates [3-6].

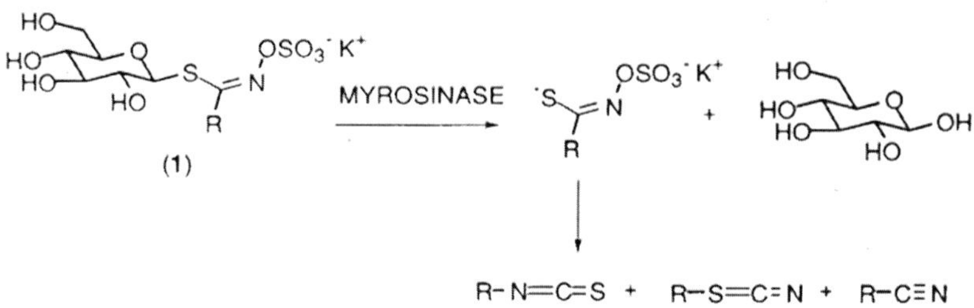

Figure 8. Myrosinase-enzyme catalyzed reaction of glucosinolates ($R=CH_3SO(CH_2)n$-) or ($[CH_3SO_2(CH_2)n]$.

Characterization of O-Ethyl-4-[α-L-rhamnosyloxy benzyl]-carbamate

The mass spectral peak at *m/z* 341, present in *Moringa oleifera* pulp (Figure 9), was further analyzed by mass spectrometry MS2 technique and has been rationalized as O-ethyl-4-[α-L-rhamnosyloxy benzyl]-carbamate (Figure 9). Its structure was further corroborated by the loss of a characteristic mass fragment at *m/z* 179 corresponding to benzyl carbamate. Guevara and coworkers (*19*) have reported the presence of such a compound in *Moringa oleifera* Lam as an antitumor promoter.

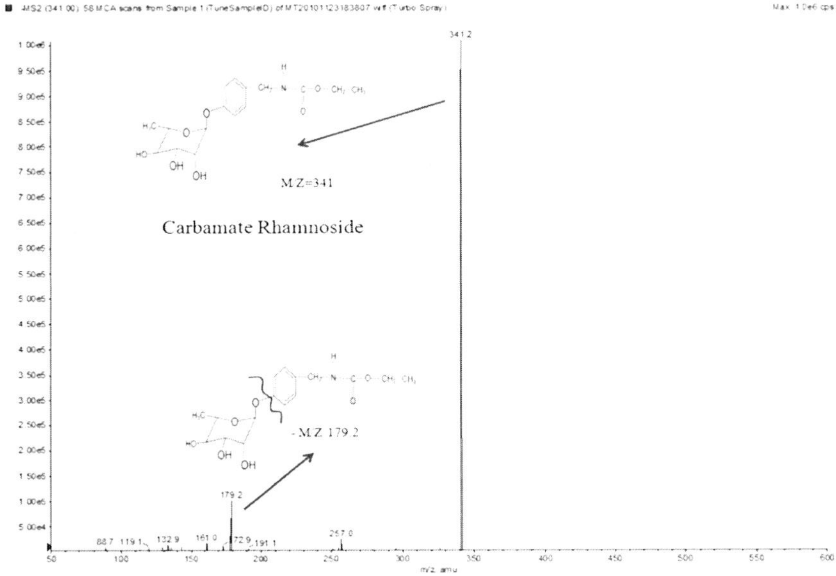

Figure 9. ESI-MS (-ve ion) of O-ethyl-4-[α-L-rhamnosyloxy benzyl]-carbamate from Moringa oleifera seedcoat/(inner skin skeleton).

The mass spectral peak at *m/z* 191 (R_f 0.94) and R_f sulforaphane standard 0.98 (Figure 7 Compound 3), present in *Moringa oleifera*, has also been analyzed using the MS2 technique and has been rationalized as 5-methylsulfinylpentylisothiocyanate (Figure 7 Compound 6). Its structure was further confirmed by the loss of isothiocyanate (*m/z* 58) (–N=C=S) to give rise to a fragment at *m/z* 133, as illustrated in (Figure 5).

Inhibitory Activity of Colon Cancer Cell Proliferation

Antiproliferative activities of all fractions and pure compounds isolated from *Moringa oleifera* were measured by Caco-2 and HT-29 human colon cancer cell lines. Figures 9 and 10 illustrate the inhibition of growth of Caco-2 and HT-29 human colon cancer cells after incubation for 72 hours with fractions from pulp, seed coat and seeds of *Moringa oleifera.* The cell proliferation was assessed by a tetrazolium salt reduction assay as previously described (*15*, *39*, *40*). Briefly, the cells were seeded on a 96-well plate at a density of $5\text{x}10^4$ in 0.2 ml medium and allowed to adhere for 24 hours. Subsequently, the cells were incubated with *Moringa oleifera* extracts. After 72 hours, 10 µL of WST-1 reagent was added to each well. After a further 1-4 hours, absorbance was measured at 450 nm on a microplate reader (Bio-Tek KC-4).

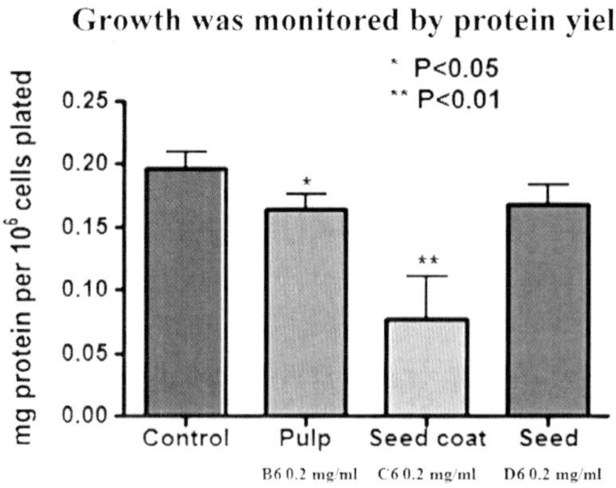

Figure 10. Inhibition of growth of Caco-2 human colon cancer cells after incubation for 72 hours with fractions from inner skin skeleton/pulp (B6), seed coat (C6) and seeds (D6) of Moringa oleifera. Designation: A6 (skin), inner skin skeleton/pulp (B6), seed coat (C6) and seeds (D6).

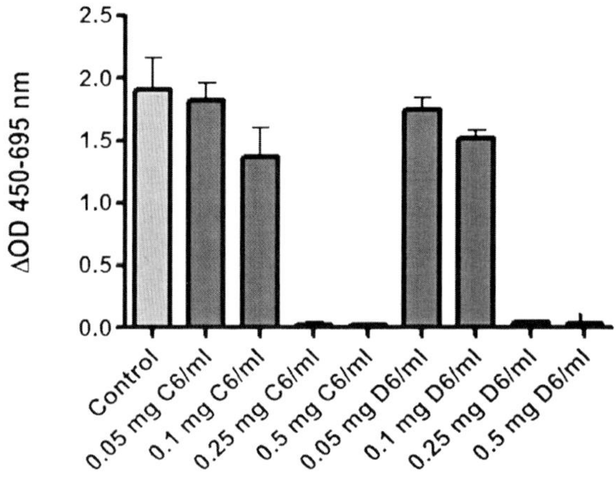

Figure 11. Inhibition of cell proliferation by seed coat and seeds (D6 fractions of Moringa oleifera after incubation for 72 hours with HT-29 human colon cancer cells. Proliferation was monitored by a tetrazolium salt reduction assay (A6 skin, B6 inner skin skeleton/pulp, C6 seed coat, D6 seed).

Our studies demonstrate, for the first time a systematic structure elucidation and the bioactivities of *Moringa oleifera* specifically the comparative antiproliferative activities in a dose dependent manner [*50μg/mL, 100μg/mL, 250μg/mL* concentrations of pulp (or inner skin skeleton), seed coat and the seed]. (Figures 10 and 11).

Isolated fractions of pulp, seed coat and the seed (0.2 mg/ml) inhibited the growth of Caco-2 human colon cancer cells after 72 hours of incubation (Figure 9). A comparative inhibition of cell proliferation by fractions from seed coat and seeds of *Moringa oleifera* (0.25mg/ml and 0.5 mg/ml) after 72 hours of incubation with

HT-29 Human colon cancer cells is shown in Figure 10. Induction of apoptosis and antiproliferative properties from *Moringa* oleifera *leaf extract* on human cancer cells are in line with previous findings as elucidated by Sreelatha and coworkers (*41*). We believe that the antiproliferative activities of pure fractions from *Moringa oleifera* may have beneficial effects for the prevention and treatment of diabetes, cancer and various neurodegenerative diseases.

Recent studies by Mahajan et al. reported that Moringa oleifera roots inhibited tumor necrosis factor-α (TNFα) and IL-2 while seeds decreased the concentration of IL-6 (*21*). These are very important observations as these compounds could serve as potential new medicines. Studies from our laboratory have indicated that IL-6 accelerates high-sensitivity C-Reactive Protein pathway, (hs -CRP a cardiovascular biomarker) (*49*). It must be pointed out that IL-6 and hs-CRP are both cardiovascular biomarkers and will be discussed in detail in future studies (*50*).

Conclusions

The present studies and the work done by other investigators provide convincing data for many valuable medicinal uses of this plant and needs to be explored further to achieve lead molecules for novel herbal drugs. The hydrolysis of glucosinolates by the plant enzyme myrosinase generates isothiocyanates which exert anticancer effects. Seeds of this plant decrease the concentration of IL-6, a cardiovascular biomarker and may serve as potential new medicines.

Acknowledgments

We are grateful to Jason Mahr, a summer research scholar in the department of biochemistry and molecular biology, for his excellent technical laboratory experimentation and skillful computer editing expertise in making this manuscript possible. We are also thankful for his help in preparing the excel diagrams and figures for this manuscript. The help of Cadisha Becket and Jironia Noel is also acknowledged.

References

1. Fahey, J. W. *Moringa oleifera*: A review of the medical evidence for its nutritional, therapeutic, and prophylactic properties. Part 1. *Trees Life J.* **2005**, 1–5.
2. Anwar, F.; Latif, S.; Ashraf, M.; Gilani, A. H. *Moringa oleifera*: A food plant with multiple medicinal uses. *Phytother Res.* **2007**, *21*, 17–25.
3. Posmontier, B. The medicinal qualities of *Moringa oleifera*. *Holist. Nurs. Pract.* **2011**, *25* (2), 80–87.
4. Murakami, A.; Kitazono, Y.; Suratwadee, J.; Koshmizu, K.; Ohigashi, H. Niazimicin, a thiocarbamate from leaves of *Moringa oleifera*, holds strict structural requirement for inhibition of tumor-promoter-induced Epstein-Barr virus activation. *Planta Med.* **1988**, *64*, 319–323.

5. Fakurazi, S.; Hairuszah, I.; Nanthini, U. *Moringa oleifera* Lam prevents acetaminophen-induced liver injury through restoration of glutathione level. *Food Chem. Toxicol.* **2008**, *46* (8), 2611–2615.
6. Cheenpracha, S.; Park, E. J.; Yoshida, W. Y.; Barit, C.; Wall, M.; Pezzuto, J. M.; Leng, C. C. Potential anti-inflammatory phenolic glycosides from the medicinal plant *Moringa oleifera* fruits. *Biorg. Med. Chem.* **2010**, *18*, 6598–6602.
7. Faizi, S.; Siddiqui, B. S.; Saleem, R.; Noor, F.; Husnain, S. Isolation and structure elucidation of a novel glycoside niazirin from the pods of *Moringa oleifera*. *J. Nat. Prod.* **1977**, *60*, 1317–1321.
8. Faizi, S.; Siddiqui, B.; Saleem, R.; Siddiqui, S.; Aftab, K.; Gilani, A Novel hypotensive agents, niazimin A, niazimin B, niazicin A and niazicin B from *Moringa oleifera*: Isolation of first naturally occurring carbamates. *J. Chem. Soc.* **1994**, *1*, 3035–3640.
9. Gueyrard, D.; Barillari, J.; Iori, R.; Palmeri, S.; Rollin, P. First synthesis of an O-glycosylated glucosinolate isolated from *Moringa oleifera*. *Tetrahedron Lett.* **2000**, *41*, 8307–8309.
10. Dayal, B.; Amin, R.; Yannamreddy, V.; Lea, M.; Attygalle, A. B. Breaking Up Biofilms with *Moringa oleifera*: Insights into Mechanisms. AGFD 247: 243rd American Chemical Society National Meeting, San Diego, CA, March 2012.
11. Rathi, B. S.; Bodhankar, S. L.; Baheti, A. M. Evaluation of aqueous leaves extract of *Moringa oleifera* Lam for wound healing in albino rats. *Indian J. Exp. Biol.* **2006**, *44* (11), 898–901.
12. Bharali, R.; Tabassum, J.; Azad, M. R. Chemomodulatory effect of *Moringa oleifera* Lam on hepatic carcinogen metabolishing enzymes antioxidant parameters and skin papillomagensis in mice. *Asian Pac. J. Cancer Prev.* **2003**, *51* (8), 131–139.
13. Chuang, P. H.; Lee, C. W.; Chou, J. Y.; Murugan, M.; Sheih, B. J.; Chen, H. M. Anti-fungal activity of crude extracts and essential oil of *Moringa oleifera* Lam. *Bioresour. Technol.* **2007**, *98* (1), 232–236.
14. Sashidara, K. V.; Rosaiah, J. N.; Tyagi, E.; Shukla, R.; Raghubir, R.; Rajendran, S. M. Rare dipeptide and urea derivates from roots of *Moringa oleifera* as potential anti-inflammatory and antinciceptive agents. *Eur. J. Med. Chem.* **2009**, *44* (1), 432–436.
15. Lea, M. A.; Akinpelu, T.; Amin, R.; Yarlagadda, K.; Enugala, R.; Dayal, B.; desBordes, C. Inhibition of growth and induction of differentiation of colon cancer cells by extracts from okra (*Abelmoschus esculentus*) and drumstick (*Moringa oleifera*). *Proc. Am. Assoc. Cancer Res.* **2012**, *53*, 1314–1315.
16. Ndabigengeserea, A.; Narasiah, K. S. Quality of water treated by coagulation using *Moringa oleifera* seeds. *Helv. Chim. Acta* **1998**, *32*, 81–791.
17. Rashid, U.; Anwar, F.; Moser, B. R.; Knothe, G. *Moringa oleifera* oil: A possible source of biodiesel. *Bioresour. Technol.* **2008**, *99*, 8175–8179.
18. Park, E. J.; Cheenpracha, S.; Chang, L. C.; Kondratyuk, T. P.; Pezzuto, J. M. Inhibition of lipopolysaccharide-induced cyclooxygenase-2 and inducible nitric oxide synthase expression by 4-[(2′-O-acetyl-α-L-rhamnosyloxy)

benzyl] isothiocyanate from *Moringa oleifera*. *Nutr. Cancer* **2011**, *63*, 971–982.
19. Guevara, A. P.; Vargas, C.; Sakurai, H.; Fujiwara, Y.; Hashimoto, K.; et al. An antitumor promoter from *Moringa oleifera* Lam. *Mutat. Res.* **1999**, *440*, 181–188.
20. Fahey, J. W.; Zalcmann, A. T.; Talalay, P. The chemical diversity and distribution of glucosinolates and isothiocyanates among plants. *Phytochemistry* **2011**, *56*, 5–51.
21. Mahajan, S. G.; Mali, R. G.; Mehta, A. A. Protective effect of ethanolic extracts of seeds of *Moringa oleifera* Lam against inflammation associated with development of arthritis in rats. *J. Immunotoxicol.* **2007**, *4* (1), 39–47.
22. Amaglo, N. K.; Bennett, R. N.; Lo Curto, R. B.; Rosa, E. A. S.; Lo Turco, V.; Giuffrida, A.; Lo Curto, A.; Crea, F.; Timpo, G. M. Profiling selected phytochemicals and nutrients in different tissues of the multipurpose tree *Moringa oleifera* L., grown in Ghana. *Food Chem.* **2010**, *122*, 1047–1054.
23. Verma, A. R.; Vijayakumar, M.; Mathela, C. S.; Rao, C. V. *In vitro* and *in vivo* antioxidant properties of different fractions of *Moringa oleifera* leaves. *Food Chem. Toxicol.* **2009**, *47*, 2196–2201.
24. Debnath, S.; Biswas, D.; Ray, K.; Guha, D. *Moringa oleifera* induced potentiation of serotonin release by 5-HT(3) receptors in experimental ulcer model. *Phytomedicine* **2011**, *18*, 91–95.
25. Xian, L.; Kushad, M. H. Correlation of glucosinolate to myrosinase activity in horseradish (*Armoracia rusticana*). *J. Agric. Food Chem.* **2004**, *52*, 6950.
26. Thurber, M. D.; Fahey, J. W. Adoption of *Moringa oleifera* to combat under nutrition viewed through the lens of the "diffusion of innovations" theory. *Ecol. Food Nutr.* **2009**, *48*, 212–225.
27. Francis, J. A.; Jayaprakasam, B.; Olson, L. K.; Nair, M. G. Insulin secretagogues from *Moringa oleifera* with cyclooxygenase enzyme and lipid peroxidation inhibitory activities. *Helv. Chem. Acta.* **2004**, *87*, 317–326.
28. Dayal, B.; Yannamreddy, V.; Racharla, L.; Desai, J.; Lea, M. Microwave-Induced Esterification of Chlorogenic Acid and Other Polyphenolic Acids Catalyzed by Hafnium (IV) chloride. AGFD 95: 241st American Chemical Society National Meeting, Anaheim, CA, March 2011.
29. Dayal, B.; Keshava, R.; Salen, G. Microwave-induced organic reactions of bile acids: Esterification, deformylation and deacetylation using mild reagents. *Steroids* **1995**, *60*, 453–457.
30. Dayal, B.; Ertel, N. H.; Rapole, K. R.; Asgaonkar, A. Rapid hydrogenation of unsaturated sterols and bile alcohols using microwaves. *Steroids* **1997**, *62*, 451–454.
31. Dayal, B; Rapole, K. R.; Salen, G. Microwave-Assisted Organic Synthesis of Sterols and Bile Acids. Orgn 46: 244th American Chemical Society National Meeting, Philadelphia, PA, August 25–29, 2012.
32. Bose, A. K.; Banik, B. K.; Lavlinskaia, N.; Jayaraman, M.; Manhas, M. S. More chemistry in a microwave. *Chemtech* **1997**, *27*, 18–24.
33. Jyothi, D.; Khanam, S.; Sultana, R. Optimization of microwave assisted extraction of Withanolides from roots of Ashwagandha and its comparison

with conventional extraction method. *Int. J. Pharm. Pharmaceut. Sci.* **2010**, *2*, 46–50.

34. Dayal, B.; Yannamreddy, V. R.; Amin, R.; Lea, M. A.; Attygalle, A. B. Bioactive Compounds Present in *Moringa oleifera*: Isolation and Structure Elucidation by Thin-Layer Chromatography and Electrospray Ionization Mass Spectrometry. 243rd American Chemical Society National Meeting, San Diego, CA, March 2012.
35. Dayal, B.; Yannamreddy, V. R.; Singh, A. P.; Lea, M.; Ertel, N. H. Bioactive Compounds from Okra Seeds: Potential Inhibitors of Advanced Glycation End Products. In *Emerging Trends in Dietary Components for Preventing and Combating Disease*; Patil, B. S., et al., Eds.; ACS Symposium Series 1093; American Chemical Society: Washington, DC, 2012.
36. Chemists Crank up Heat on Microwaves. *Chem. Eng. News* Sept. 2012, pp 32–34.
37. Bialecki, J. B.; Ruzicka, J.; Weisbecker, C. S.; Haribaland, M.; Attygalle, A. B. Collision-induced dissociation mass spectra of glucosinolate anions. *J. Mass Spectrom.* **2010**, *45*, 272–283and references cited therein.
38. Mirza, U. A.; Liu, Y.-H.; Tang, J. T.; Porter, F.; Bondoc, L.; Chen, G.; Pramanik, B. N.; Nagabhushan, T. L. Extraction and characterization of adenovirus proteins from sodium dodecylsulfate polyacrylamide gel electrophoresis by matrix-assisted laser desorption/ionization mass spectrometry. *J. Am. Soc. Mass Spectrom.* **2000**, *11*, 356–361.
39. Lea, M.; Ibeh, C.; Deutsch, J.; Hamid, I.; desBordes, C. Inhibition of growth and induction of alkaline phosphatase in colon cancer cells by flavonols and flavonol glycosides. *Anticancer Res.* **2010**, *30*, 3629–3636.
40. Lea, M.; Chacko, J.; Bolikal, S.; Hong, J.; Chung, R.; Ortega, A.; desBordes, C. Addition of 2-deoxyglucose enhances growth inhibition but reverses acidification in colon cancer cells treated with phenformin. *Anticancer Res.* **2011**, *31*, 421–426.
41. Sreelatha, S.; Jeyachitra, A.; Padma, P. R. Antiproferative and induction of apoptosis by *Moringa oleifera* leaf extract on human cancer cells. *Food Chem. Toxicol.* **2011**, *49*, 1270–1275.
42. Bennett, R. N.; Mellon, F. A.; Foidl, N.; Pratt, J. H.; Dupont, M. S.; Perkins, L.; Kroon, P. A. Profiling glucosinolates and phenolics in vegetative and reproductive tissues of the multi-purpose trees *Moringa oleifera* L. (Horseradish Tree) and *Moringa stenopetala* L. *J. Agric. Food Chem.* **2003**, *51* (12), 3546–3553.
43. Botting, C. H.; Hanschen, F. S.; Rohn, S.; Mewis, I.; Schreiner, M.; Kroh, L. W. Influence of the chemical structure on the thermal degradation of the glucosinolates in broccoli sprouts. *Food Chem.* **2012**, *130*, 1–8.
44. Ganin, H.; Rayo, J.; Amara, N.; Levy, N.; Krief, P.; Meijler, M. M. Sulforaphane and erucin, natural isothiocyanates from broccoli, inhibit bacterial quorum sensing. *Med. Chem. Commun.* **2013**, *4*, 175–179.
45. Zhang, Y.; Talalay, P.; Cho, C. G.; Posner, G. H. A major inducer of anticarcinogenic protective enzymes from broccoli: Isolation and elucidation of structure. *Proc. Natl. Acad. Sci.* **1992**, *89*, 2399–2403.

46. Lee, Y. M.; Seon, M. R.; Cho, H. J.; Kim, J. S.; Park, J. H. Benzyl isothiocyanate exhibits anti-inflammatory effects in murine macrophages and in mouse skin. *J. Mol. Med.* **2009**, *87*, 1251–1261.
47. Juge, N.; Mithen, R. F.; Traka, M. Molecular basis for chemoprevention by sulforaphane: A comprehensive review. *Cell. Mol. Life Sci.* **2007**, *64*, 1105–1127.
48. Leuck, U.; Kunz, H. Synthesis of active principles from the leaves of *Moringa oleifera* using S-pent-4-enyl thioglucosides. *Carbohydr. Res.* **1998**, *312*, 33–44.
49. Dayal, B.; Ertel, N. H. ProteinChip technology: A new and facile method for the identification and measurement of high-density lipoproteins apoA-I and apoA-II and their glycosylated products in patients with diabetes and cardiovascular disease. *J. Proteome Res.* **2002**, *1* (4), 375–380.
50. Dayal, B.; Ertel, N. H. Cardiac Molecular Markers (II): Detection and identification of hs-CRP and Interlukin-6. Activation of High Sensitivity C-Reactive Protein in Patients with Diabetes and Cardiovascular Disease by SELDI ProteinChip Arrays. Presented in part at the American Chemical Society National Meeting, (abstr. 177, Agricultural and Food Chemistry Division , San Diego, Ca, March 13–17, 2005.

Editors' Biographies

Bhimanagouda S. Patil

Dr. Bhimu Patil is the Professor and Director of Vegetable and Fruit Improvement Center. His research focus on multi-disciplinary "Foods for Health" led to the publication of more than 135 peer reviewed research papers and 15 awards. Drs. Patil and Yves Desjardin initiated an international symposium "FAVHealth" symposium in 2005, which was hosted in Houston 2007. He has chaired or co-chaired 25 symposiums. He was an invited speaker, including plenary speaker, for his scientific research and educational excellence by several countries including China, South Korea, Brazil, Indonesia, Spain, Sweden, France, India, Canada, Portugal, and different states in the United States. He has developed two multi-disciplinary and multi-state first-of-its kind course, "Science of Foods for Health" and "Phytochemicals in Fruits and Vegetables to Improve Human Human Health".

Guddadarangavvanahally K. Jayaprakasha

G. K. Jayaprakasha has 23 years of research and 3 years of industrial experience focused on isolation of natural products from fruits, vegetables, spices, and medicinal plants as well as semi-synthesis of bioactive lead natural products and their biological evaluation for the treatment of cancer *in vitro* models. His research involves bioassay directed discovery, purification, and chemical characterization of natural compounds using preliminary *in vitro* assays such as anticancer, antimicrobial, and antioxidant properties. He has also isolated and characterized >80 novel, rare, and bioactive compounds from Citrus, Pomegranate, Carrots, Cinnamon, Lichen, Turmeric and *Garcinia*. He filed 24 patents in the United States, Europe, and India. Dr. Jayaprakasha has published >140 research papers, reviews, and book chapters, and has made 130 presentations in national and international meetings. In addition, he has co-edited one ACS symposium series book. Based on his research accomplishments, he has been admitted as "Fellow of Royal Society of Chemistry" (FRSC), Royal Society of Chemistry, Cambridge, United Kingdom.

© 2013 American Chemical Society

Coralia Osorio Roa

Coralia Osorio Roa, Dr. Sc, belongs to the Department of Chemistry of Universidad Nacional de Colombia-sede Bogotá, as an Associate Professor. Her research program is focused on sensory and biofunctional properties of foods and related chemistry. Particular areas of emphasis include flavor chemistry, pigments in foods (anthocyanins and carotenoids), natural product chemistry, and microencapsulation by spray-drying. Recent efforts have been moved to the chemical ecology field. During this time, she has published over 40 original research articles and some book chapters.

Kanjana Mahattanatawee

A Thai National, Kanjana Mahattanatawee, received a B.Sc. in Microbiology from Sri-Nakharinwirot University in 1988 and a M.Sc. in Industrial Microbiology from Chulalongkorn University in 1991, both in Bangkok, Thailand. In 1993 after an UNESCO fellowship in Biotechnology at Osaka University in Japan, she returned home to be a lecturer at Department of Food Technology, Siam University, Bangkok. In 2004, she finished her doctoral study in Food Science (Flavor Chemistry) under the tutelage of Prof. Russell Rouseff at the University of Florida. After a two years postdoctoral fellowship conducting research in tropical fruit flavor and bioactive compounds at USDA, ARS, Winter Haven, Florida, she returned home to Siam University. Currently, she is the Dean of the Faculty of Science, Siam University. She is also a former chair of the ACS Division of Agricultural and Food Chemistry Flavor Sub-Division, a regular reviewer for the ACS Journal of Agricultural and Food Chemistry, and the Chair of the ACS International Chemical Science Chapter in Thailand (ACS-ICSCT).

Indexes

Author Index

Subject Index

H

I

K

M

N

P

R

S

T

U

V

W